T0335260

THE SNAKES OF ONTARIO

PLATE I: Eastern Garter Snake

THE

SNAKES OF ONTARIO

E. B. S. Logier

ROYAL ONTARIO MUSEUM
DIVISION OF ZOOLOGY AND PALAEONTOLOGY

Illustrated by the Author

UNIVERSITY OF TORONTO PRESS

Copyright ©, Canada, 1958, by
University of Toronto Press

Reprinted 1967, 2017

ISBN 978-1-4875-8684-3 (paper)

INTRODUCTION

ALTHOUGH THIS LITTLE BOOK was written primarily as a guide to the snakes of Ontario, it does, in fact, deal with all the snakes of eastern Canada, with the exception of the Black Racer, which has been reported only from the Maritimes. All the other eastern Canadian snakes are found in Ontario. The discussions of distribution, however, are confined to Ontario. For each species (or race) two maps are given, one showing its range in Ontario, and the other its entire range in North America. It will be seen from these maps that the Ontario populations are extensions of those in the United States, and that if considered alone—that is, without reference to the main bodies of the populations to the south, west, or east—they may not seem to make sense.

Since it is intended mainly for young readers, the matter has been presented in an easy, conversational manner, especially the general account of snakes in the first section of the book. The second section, dealing with the separate species, did not lend itself as freely to that style of treatment, but a special effort was made to keep it simple, non-technical, and easily readable. Only the English names of the snakes have been used in the text and headings, and technical words have been avoided almost completely.

It would not be possible, or desirable, in a book of this kind to cite the many authorities to whom I am indebted for much of the information set down in the following pages. However, a short reading list is given, and those consulting the books in it will find references to a wide field of writings on snakes and other reptiles.

For the benefit of those desiring to pursue the study of snakes more fully, the following information has been added, in appendixes: a list of the snakes of Ontario giving their scientific names; an explanation of the purpose of the Latinized scientific names; a key for the identification of Ontario snakes; directions for determining the sex of snakes; directions for collecting and preserving snakes; directions for keeping them as pets; some diagrams giving the anatomical names of the scales, and other features; a glossary explaining the meanings of words which may be unfamiliar to young readers, or to others unacquainted with the subject; and finally, a short discussion of rattlesnake bite and first aid treatment.

When the publishers requested me to address this book to young readers it gave me considerable pleasure and satisfaction. It has long been apparent that any attempt to educate the public into reasonable thinking about snakes, if it is to be successful, must start with the children, whose minds are still open to receive the truth. It is a thankless task to try to teach people the truth about snakes, or anything else, after its rightful place in their minds has been firmly preoccupied by preferred ideas. This stubborn fact, which has always slowed the advancement of knowledge and sound thinking, is largely responsible for the tenacity of error, and the persistence of much of the rationalized nonsense that people believe.

We read in a source of very wise counsel that "the truth shall make you free," but it cannot do that unless your mind is open to receive it. This is as true about your attitude toward snakes as it is about everything else, and many people live out their lives in a senseless slavery of fear of snakes because their minds were filled with nonsense in their childhood, when they should have been learning the truth.

In conclusion I should like to express my sincere thanks to the following: Dr. F. A. Urquhart, Head of the Division of Zoology and Palaeontology of the Royal Ontario Museum, for his advice and encouragement in the planning of this book; Dr. W. G. Brown, Chief Medical Officer of Health for Ontario, and Dr. R. I. Harris, F.R.C.S., for giving valuable criticism and advice on the section on First Aid, and Dr. Harris also for acting as medical consultant on the section addressed to the attending physician and hospital staff; Mr. C. D. Weber, of John Wyeth and Brother (Canada) Limited, for supplying details concerning antivenin, and other pertinent information; Mr. Hugh Halliday for the splendid photographs of the rattlesnakes (pp. 59, 61); Mr. T. M. Shortt, Senior Artist in the museum, for his valuable counsel in preparing the illustrations; Miss Eileen McClure, Librarian in the Division of Zoology and Palaeontology, for reading the proofs, and rendering other assistance.

Finally, grateful acknowledgment is due my wife for her helpful comments, and for her good-natured endurance of my absence from much of the normal activity of family life during the months of close application to meet the deadline.   E.B.S.L.

CONTENTS

ILLUSTRATIONS

Line Drawings

Maps

A distribution map is provided for each snake

THE SNAKES OF ONTARIO

What are Snakes?

YOU CAN EASILY SEE a lot of difference between a trout and a turtle, a sparrow and a bullfrog, a garter snake and a dog, and would have no difficulty in distinguishing any of these from an earthworm, a centipede or a beetle. But you might think that the snake resembled the earthworm more closely than it did the dog. The fish, frog, turtle, snake, sparrow and dog, however, have in common one important structure which the other three do not: they all have a backbone or spine and, therefore, are known as *vertebrate* animals. Animals without a backbone, such as the earthworm, centipede and beetle, are called *invertebrate* animals.

The vertebrate animals are divided into five main classes: fish, amphibians, reptiles, birds, and mammals. The members of the first three classes are called "cold-blooded" because they do not keep up a constant degree of body heat, as do the last two, which are called "warm-blooded." This is important, and we will come back to it later.

Snakes, along with turtles and lizards, belong to the reptiles, the highest class of cold-blooded vertebrates; so they are not the lowly, primitive creatures that many people believe them to be (a sort of worm with a head on it). They are also the most specialized, specially modified to meet particular conditions of life, as well as the most successful and widely distributed branch of their class. Snakes of one kind or another are found in all warm and temperate continental lands of the world, and on many of the islands of the seas. There are species of snakes adapted to every kind of habitat (living place) from deserts to rain forests and from high mountain slopes down to, and even into, the sea. Some live in trees, others burrow in the ground; most kinds live mainly on the surface of the ground, but many enter the waters of lakes and streams where they swim and dive with graceful ease.

Their greatly elongated, limbless bodies were developed along with, and fitting them for, special ways of life. A snake's body, like that of other reptiles, is covered with a dry, scaly skin. It is not slimy. The scales vary in size and arrangement on different parts of the body. The scales of the sides and back, called *dorsal scales*, are small and diamond-shaped; those of the belly are usually broadened into transverse (crosswise to the body length) plates called *ventrals*; those of the head vary in size and shape and are named according to their position. In case you wish to learn more about the scales, the diagram on Fig. 1, p. 70 shows their names and positions. It is helpful to learn to know the names of the various scales, because they are referred to by name in most good books on snakes.

How Do Snakes Travel?

IF YOU LOST your arms and legs, you would be very helpless indeed. You could not walk or swim, or defend yourself from attack, or do most of the things that you now do quite easily and naturally. But if you had a long, flexible, muscular body like a snake's, you would not be as helpless. The next time you find a snake, don't kill it or run away. It will probably try to run away from you; so just wait and see how it does this. As it glides away, you will notice that it throws its body sideways into a series of waves which flow from the head to the tail, just as an eel does in swimming. The result is the same: the waves in the eel's body push backward against the water, the waves in the snake's body push backward against the ground; both animals propel themselves forward thus. When crawling along slowly, the snake uses its ventral plates, referred to above. These are attached to muscles and can be moved backward and forward. The backward and forward movement of the ventrals progresses as a series of waves along the snake's belly, while the back edges of these plates push against the ground, moving the snake forward. A snake swims like an eel, except that the eel has the added advantage of fins. However, a snake swims very well indeed without them. In some snakes that live in the water the tail, or even the backward half of the body, is flattened on the sides, thus increasing the power of the wave strokes. When climbing, a snake uses the muscular grip of its body, as well as the motion of its ventral plates. Many snakes are good climbers, and some can even glide with their swimming-like motion over the top twigs of bushes.

2

Senses

YOUR SENSES, sight, hearing, "feeling," taste and smell, tell you what is going on in the world around you, and you depend upon them for all your enjoyment of life. The sense of "feeling" as used above includes several distinct senses, as touch, pain, heat and cold. Without your senses you would find life very dull if, indeed, you could manage to live at all. If you could not see or hear, you could not enjoy a hockey game or a television show. If you could not taste or smell, you could not enjoy your Christmas dinner; you would get no more fun out of eating your turkey and pudding than eating boiled sawdust.

Other vertebrate animals, including snakes, are more or less like you in the importance of their senses, but they do not always have the same senses in the same degree, and some seem to have other senses that you have not. When a particular sense is weak or absent, some other is often especially well developed to make up for it.

SIGHT. Most snakes have good eyesight for seeing near objects and perceiving movement, but it is not known if, or how well, they can see colour. Those with round eye pupils (Figure 2A and B) can, like you, see only in fairly good light; those with vertical eye pupils, like a cat's (Figure 3B), can see also at night. Perhaps I should explain at this point that no animal can truly "see in the dark," as people often wrongly believe about such creatures as cats and owls. For an animal to see, there must be some light; it may be so faint that human eyes may fail to detect its presence, but the sensitive eyes of nocturnal (night-roaming) animals can use it.

The eyes of snakes are made differently from those of most vertebrates. They have no movable eyelids. The lower eyelids have become transparent scales, permanently sealed over the eyes, which therefore appear to be always open and staring. Actually, they are always closed. There are other important differences too.

HEARING. Snakes have very poor hearing: experiments have shown that they are deaf to the ordinary sounds carried through air which you hear so easily, but are sensitive to sounds conveyed through solid objects, or the ground, on which they are resting. This would restrict them to low-pitched sounds. You may have noticed that the very deep notes of a church organ can be felt as vibrations in the building as well as being heard by the ear; a loud peal of thunder or the firing of a cannon can also be felt as well as heard, and often makes windows rattle. Snakes would hear those kinds of sounds, but whether they would be aware of

3

them as sounds, as you know them, or merely as vibrations, we do not know. They have no eardrum and no middle ear, which are so important in your hearing. It follows, then, that snakes are not "charmed" by music and will not come to the sounds of pipes or flutes.

SMELL. The sense of smell is well developed in snakes and may be nearly, or quite, as important to them as eyesight in finding their food. A snake really has two smelling organs: it has a nose that functions like yours, but it also has two little cavities at the front of the roof of its mouth (called Jacobson's organ) which are connected to the nerves of smell. When a snake "sticks its tongue out at you," it is not trying to be rude, or tell you how little it thinks of you, or even threatening to "sting" you. It is really trying to find out what sort of a creature you are and is doing the same kind of thing, by a different method, that your pet dog does when he meets a stranger: just using its sense of smell. A snake's tongue is not a "stinger," it is soft and fleshy, and is really part of the smelling system. Since the tongue is wet, when it is thrust out and waved up and down in the air, or made to touch anything, odours are picked up by it and cling to the two little forks (Figure 3D); when it is withdrawn into the mouth the odours are brushed off into the two little cavities mentioned above, where they are received as smells. The tongue itself, so far as we know, has no sensory function, such as taste, or hearing, and is used only as described above. However, snakes depend upon it a lot and are constantly flicking it out when crawling about or hunting their prey.

OTHER SENSES. I do not know whether snakes can taste or not. They have been known to swallow such inedible objects as stone or china nest eggs from domestic hen nests, which would, no doubt, have an attractive odour of the birds. If we include flavour in our definition of taste, it becomes difficult to separate it from smell. A caged python has been known to swallow its own blanket, and one may doubt that its smell would have any attraction in this case, as it might in that of the nest eggs. The other senses of snakes, such as touch, pain, heat, cold and balance, are probably quite similar to yours in the kinds of sensations they produce.

PLATE II. A, Red-bellied Snake; B, Eastern Ring-necked Snake; C, Eastern Smooth Green Snake

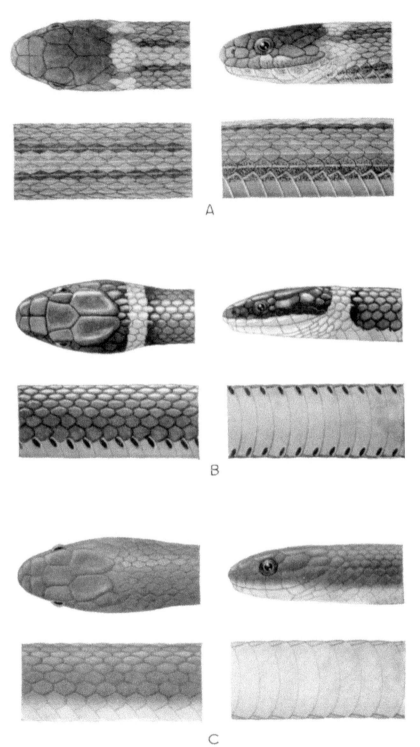

A

B

C

Instinct and Intelligence

SNAKES ARE NOT very intelligent. Perhaps you could truthfully say the same thing about some of the people you know, but it is usually better not to. All vertebrate animals have some measure of intelligence, but the lower down you go in the scale, the less you find. The terms "lower" or "higher" are generally applied to vertebrates from the human viewpoint, that man is the highest creature, which, however, is strictly true only with regard to his intelligence and his ability to use his hands. In other respects many of the "lower" animals are more highly perfected for their special ways of life. Intelligence is manifested by ability to think, learn and remember, and to meet new kinds of situations and solve new problems.

If you have kept pets, and among them were dogs, cats, turtles and snakes, you will have noticed a great difference in how much and how quickly these different animals learned and responded to you. Your dog soon learned his name, learned the meaning of some words, your tone of voice, and to obey your commands (except when disobedience was much more attractive). He invented his own ways of telling you what he wanted, and liked to engage you in play. Your cat displayed some of these powers, but to a lesser degree than your dog. With your snakes and turtles the story was rather different. After a few months they lost much of their fear of you and learned to associate your presence with feeding, so that if hungry they might come to your hand and start nosing around. Perhaps they even became tame enough to seem to enjoy a little gentle handling and petting, but at that point your snakes' learning seemed to end; perhaps some of your turtles went a little further.

In lower animals such as snakes and turtles, intelligence is almost entirely replaced by another faculty, instinct, which directs the lives of these animals. Instinct might perhaps be described as an inherited impulse to behave in a particular way under particular circumstances. It is a sort of "unconscious knowing" without learning. The young snake does not have to learn where or how to hunt its prey, or what kind of prey to hunt, or how to defend itself; if it is a water snake, it does not have to learn to swim and catch fish; if it is a milk snake, it does not have to learn how to kill its prey by constriction. When its first autumn comes, it does not have to learn where to hibernate for the winter, and when it becomes an adult, it does not have to learn where to find its mate, or how to recognize it, or where to lay its eggs. Under natural conditions it does the right thing

5

at the right time; so it is not necessary for a young snake to learn, it grows up with the sense with which it was born.

Nobody knows what instinct and intelligence really are, just where or how the one grades into the other, or how they are related to the brain and nervous system of the creatures that use them. That an animal lacks intelligence, as we understand it in the human sense, is no reason for holding it in disrespect. The instinct which directs the lives of such animals is one of the most wonderful things in the living world, and without it the living world would have ended long, long ago, and even man would not be here. However, it has its limitations, and may be unable to act "wisely" in new situations, but will faithfully repeat the accustomed form of behaviour, whether it is appropriate to the new situation or not. Intelligence is superior in that it can "size up" a situation and modify behaviour to meet it.

How Snakes Feed

ALL SNAKES ARE carnivorous, that is to say, they eat other animals, but they do not all eat the same kinds of animals. Some snakes, such as the garter snake, will eat almost any kind of animal small enough to swallow, but some other species are more restricted or "specialized" in their feeding. The hog-nosed snake, for instance, lives on toads and frogs; the smooth green snake feeds chiefly upon grasshoppers, caterpillars and spiders; the milk snake eats lizards, snakes, small mammals and birds; the water snake lives on fish, frogs and salamanders. Some snakes will occasionally eat animals that do not ordinarily belong to their diet, as when a water snake eats a mouse. Many snakes will eat fresh, dead animals, others will accept them even after they have begun to decay.

KILLING OF PREY. Without limbs and claws, snakes would appear to be at a disadvantage; how then do they catch and overpower their often swift and agile prey? First, they stalk it by stealth, or lie in wait until it comes near, and when it is close enough, strike quickly. The prey is overpowered by one of four means: (1) it is held fast by the sharp teeth and swallowed alive; (2) it is subdued by pressure of the snake's body against some rigid object; (3) it is constricted in coils of the snake's

6

body until struggling has ceased; (4) it is killed or paralysed by venom injected when the snake strikes.

SWALLOWING. Eating food is, for you, a pleasant and simple procedure. You bite (break or cut) off a piece at a time that will fit comfortably into your mouth, then you chew it for a few moments, and swallow it without effort. But if you had no hands, and your teeth could not bite off or chew, and you wanted to eat a melon or a turkey, you would be in rather a fix. Eating would then be a very difficult problem, and unless you could get soft, mushy foods that needed no chewing you would soon starve.

Snakes today are in exactly the same position in so far as tearing, biting or chewing their food goes, but they met this problem in a very interesting way, and they did not have to meet it suddenly. Their swallowing mechanism was developed slowly through a long biological history to meet the situation which gradually developed along with it. The tooth-bearing bones of the upper jaw are not rigidly attached to the skull, as yours are, but are movable. The lower jaw bones of each side are united in front by an elastic ligament, and attached to the skull by a chain of movable bones. The teeth are sharp, needle-shaped spikes with their points directed backward. The skin of the throat, neck, and body is capable of great stretching, and the ribs can spread outward. This elastic arrangement allows a snake to stretch its mouth over, enclose, and finally swallow, an object much larger than its own head.

Swallowing is accomplished by a "walking" movement of the jaws over the food, those of each side alternately being pushed forward and then drawn backward, while the sharp, backward-pointing teeth only grip on the backward pull, forcing the food down the throat, where the body muscles work it down to the stomach. The stomach itself is very stretchable and capable of receiving a surprisingly large bulk of food.

When you swallow, your glottis (the opening into your wind-pipe) closes for a moment, but for you swallowing does not last long enough to really interfere with your breathing. But with a snake, swallowing may take from ten or fifteen minutes to an hour or more; so it must be able to breathe while swallowing. This is accomplished by the special construction of the glottis (Figure 3D), which can be forced forward in the floor of the mouth until it sticks out in front, and then opens and allows the snake to breathe. This is repeated as often as necessary during a long swallow.

The first time I was bitten by a milk snake, which grabbed my finger after I had picked it up, I noticed two very interesting things. First, there was no pain; the bite did not hurt until I tried to pull my finger away. Second, I could not pull my finger out of the snake's seemingly feeble jaws against the grip of the tiny, sharp, backward-pointing teeth without tearing the skin. In order to release it, I had to push my finger toward the snake's throat, slipping it out sideways at the same time. I realized then how very securely a small snake can hold its struggling and sometimes rather large prey.

DIGESTION. Snakes digest their food slowly, and whenever possible they eat large meals, often as much as 20 or 25 per cent of their own weight. They use up very little energy since they have no constant body heat to maintain, as humans and other warm-blooded animals have, and when not hunting, they live very inactive lives. One good meal may serve the actual bodily requirements of a snake for a month, or several months, and may contain more calories than it would use in a year. However, food is not retained in the stomach and intestines until all its energy has been extracted, and snakes feed much oftener.

Reproduction

MOST SNAKES LAY EGGS, but quite a number give birth to active young. What usually happens in the latter case is, apparently, that the eggs are simply retained inside the body of the mother snake until the young are hatched. There is a great advantage to this in that the developing young are protected from enemies, and from unfavourable conditions such as wet or drought, or too much heat or cold. The eggs are fairly safe in the secluded nests where the snakes hide them, in rotting logs, under stones, in decaying vegetation, etc., but are not as safe as in the body of the mother, who seeks comfortable temperatures, escapes or fights off enemies, and moves away from floods and drought. Unless the snake dies, or is injured in a serious way, the conditions inside her body will be kept exactly right for the developing young.

Some snakes take care of their eggs while they are hatching, but most species merely deposit them in a proper situation and leave them. So far as I know, no snakes take care of their young after they are born or hatched. They are left to shift for themselves and are able to do so very well.

The numbers of eggs or young varies with the species. Our ring-necked snake usually lays only three eggs; the smooth green snake lays six or seven; a blue racer may lay twenty-five. The red-bellied snake usually produces seven or eight young—rarely, as many as thirteen; the northern water snake usually from about twenty to occasionally forty; the eastern garter snake usually from about twelve to thirty, but as many as seventy-eight have been recorded in one litter. So, you see, you cannot say how many children a snake ought to have, even if you know the species. When four or five babies are born together into a human family, the event is so unusual that the world goes mad about it, but most snakes would consider it a pretty poor showing.

Snakes in temperate climates, such as that of Ontario, mate in the spring very soon after awakening from their winter's sleep; some mate in the late summer also, but the young from late matings are not produced until the following year. After mating, it takes several months for the young to develop to the point where they may be hatched or born, and this must happen well in advance of the onset of autumnal coolness. So whether the young are hatched from eggs or born active, they usually appear in August or September. They must have time to make preparation for the winter.

Nearly everything living on dry land has to make some preparation for winter. People get their warm clothes out of storage and put up their storm windows and, perhaps, lay in a stock of coal. Squirrels gather a store of nuts and grow warmer coats. Many caterpillars spin cocoons. Snakes do none of these things. They simply eat until they put on enough fat to tide them over their winter's sleep, but since they can only do this on warm days, the young ones must get started early. Nature thus co-ordinates the necessary activities of living things so that each step is taken at the right time, and with sufficient time for the next one that must follow. This timing is one of the truly great marvels of the "unconscious knowing" of living creatures. "Too little and too late" is a human failing in which man's judgment proves to be at fault, or his laziness prevents him acting in time.

Hibernation

ON THE FIRST PAGE of this book I mentioned the fact that reptiles are cold-blooded, meaning that their body temperature is not maintained at an even level, but varies with that of their surroundings.

Your body stays at a temperature of 98.6° Fahrenheit, no matter what the weather, and if your temperature were to drop by only a few degrees, you would become unconscious. In the winter you eat high calorie foods—foods that supply heat—you heat your house, and put on a warm coat when you go out. When exposed to too much cold, you shrink away from it, feel very uncomfortable, and may even shiver (which is an involuntary effort of your muscles to produce more heat). Because of man's need of warmth, the words "cold" and "chilly" have become symbols in our language for disagreeable relationships: you may say that you had a "chilly" reception, or that your friend was "cold" to you.

A snake's reaction to cold is different from yours. Its body does not put up a fight to keep its temperature at the preferred level. When it is comfortably warm, it will move away from a colder temperature if possible, but if exposed to a gradually falling temperature, its body slowly cools down without resisting. A snake, therefore, does not suffer from the cold as you do. As it cools off it becomes sluggish, sleepy, and finally unconscious, but freezing must be avoided.

Since the favourable temperatures for activity for snakes lie between 70° and 90° Fahrenheit, and they cannot manufacture their own heat as you do, they have to depend upon warm weather for the necessary amount of heat to enable them to feed, digest their food, grow and produce young. For this reason, in climates like Ontario's, they are forced to hibernate during the cold months. Hibernation is a very deep sleep, at lowered temperature, in which all the vital processes of their bodies are slowed down to the minimum necessary to preserve life.

In the autumn, therefore, our snakes seek sheltered holes in which to curl up and sleep throughout the winter, where they will be safe from the danger of freezing. Often many snakes will den up together in the same cavity, perhaps in an animal burrow, beneath a tree root, in an embankment, or perhaps in the masonry of a bridge or wall. Any deep, sheltered, well-drained or moderately dry place will serve their need.

If you had to fast for a month, you would lose a lot of weight, and if you did not drink for even a few days, you would die. This is because you cannot slow down the internal processes of your body, which always proceed at about the same rate. You consume a lot of energy in merely keeping your temperature at the necessary level, and evaporate a lot of water from your warm body. Therefore, you require both food and water at frequent intervals, even if you are inactive.

If you caught a snake and weighed it just before it went into hibernation in the autumn and caught and weighed it again when it awakened in the spring, you would find that although it had taken neither food nor water for about five or six months, it had lost very little weight. It would have plenty of reserve energy to mate, and hunt its first meal, and even to fast for many more weeks if food could not be found; but it must have water, its internal supply of which would soon be depleted after normal activity began.

The fact that snakes are cold-blooded (a better word is "ectothermic" meaning that they get their heat from outside their bodies) restricts them in their geographic distribution to warm and temperate lands. The further north you go (in the northern hemisphere) the fewer kinds of snakes you find. In southern Ontario there are sixteen species of snakes, but at James Bay there is only one, the eastern garter snake.

Shedding of the Skin

YOU MAY NOT KNOW this, but you are constantly shedding the outer layer of your skin in tiny flakes. The reason is that this outer layer is dead tissue, incapable of growing or replacing itself. Therefore, as it wears out, or gets too small for your growing body, it has to be replaced by living tissue from the deeper layers.

Snakes meet this condition in a different manner. Several times a year they shed the entire outer layer in one piece. Starting at the lips, the old skin splits and is peeled off inside-out, like a stocking or a closely fitting glove. The shed skin is like a nearly transparent inside-out ghost of the snake that owned it, complete in every detail, including the transparent disk-like scales that covered the eyes.

How often a snake may shed its skin depends upon several causes, not all of which are known. Most adult snakes in Ontario shed their skins three or four times a year. A young snake sheds its first skin soon after birth or hatching, often within from one to a few hours' time, and may have four or five sheddings in the first full summer of its life. Rate of growth is one of the causes of shedding, but not the only one. Warm temperature, moisture, parasites on the skin, or injury to the snake are among other causes of more frequent shedding.

About ten days before shedding, a snake's eyes become clouded and whitish, and since it is then nearly blind, it hides away beneath cover or in some sheltering crevice. Its skin is also dull in colour. About a week before shedding its eyes clear again. After shedding, the snake is brilliant in colour and pattern compared with what it was before, and is usually both hungry and lively.

Each time a rattlesnake sheds its skin it adds another segment to its rattle; so the number of segments, if the rattle is complete, will tell you how many times the snake has shed, but not its age in years. A complete rattle tapers gradually to a little rounded button at the end (the original button of the baby rattlesnake), but few adult rattlesnakes have complete rattles, because the older segments at the end get broken and worn off in time.

Defences of Snakes

MOST ANIMALS will try to defend themselves from their enemies, if they have any means of doing so. You would use your hands and feet, or some weapon if you had it, and perhaps even your teeth; but if you had no hands or feet, you would not have much of a chance for you could neither fight nor run away—even a gun would not help you then.

Snakes, without the advantage of hands or feet, have, as a group, resorted to many devices for self-defence. The snakes of Ontario make use of some of them. Most snakes, even most venomous ones, prefer escape to combat and will try first to get away and hide, fighting only if escape is cut off. Concealing colours and patterns that blend closely with the snake's normal environment (surroundings) are a common and effective protection, by making it unlikely to be seen. If at the same time the snake has the habit of "freezing" or remaining perfectly still, the invisibility may be almost as complete as if the snake were composed of gas. You can look straight at it and not see it unless it moves. This is a very common protection among many kinds of wild animals besides snakes, as any woodsman can tell you.

Venomous (poisonous) snakes, of course, are armed with venom and fangs, the deadliest means of defence in the animal world, by which they can disable or kill an enemy hundreds of times their own weight. However, most snakes have no venom or fangs, but that does not deter some of them from using their teeth, and some will bite furiously when seized. Large snakes, when they bite, can tear the skin and cause bleeding, and although no venom is injected, such bites should be treated with an antiseptic, such as iodine, because there is danger of infection.

In addition to biting, or instead of it, some snakes use a sort of chemical warfare. If grasped by the hand, or the jaws of an animal, they freely void quantities of excrement mixed with a foul-smelling secretion from the anal scent glands. Meanwhile, they thrash their bodies and tails about, smearing the unpleasant mess over themselves and their captor. This excretion is not poisonous or harmful to the skin, but it is so disagreeable and unexpected that it probably often secures the snake's release.

Bluff, a form of deception, is a common device in the human world for getting the better of a situation to which one's physical or mental resources are unequal. Enemies or competitors may be intimidated or discouraged; jobs may sometimes be obtained, and even held by it, until the bluff is "called" by something or someone that puts the bluffer's real powers to the test. Sometimes the bluff is not called, and then, from the bluffer's viewpoint, it is quite as useful as ability, but it is an inferior substitute, as anybody who thinks at all can see. While depending upon it, the bluffer fails to develop more useful and reliable faculties; so the axe of possible misfortune is always suspended above him on a hair.

Many animals make use of bluff, and snakes are among them. Many harmless snakes will vibrate their tails after the manner of rattlesnakes when frightened, and may strike and hiss at the same time. A baby milk snake about eight inches long, buzzing its tail and hissing and striking at an enemy thousands of times its size and weight, is an amusing, but also an admirable sight. You are impressed by its spirit and pluck. This little snake is using to the utmost every part and faculty it has, which is not true of the human bluffer.

Among Ontario's snakes, and perhaps among all snakes, the hog-nosed snake takes the "Oscar" for bluff. It is a totally harmless snake that cannot be induced to bite. If you surprise one in the open and cut off its retreat, it puts on a display that makes it look quite as dangerous as a cobra or a viper. It flattens its head and spreads its neck to about twice its normal width, and hissing loudly, strikes (mouth closed) at your foot or hand. If this does not frighten you away, and you hold your ground, or make passes at it, it then tries something else. Turning over on its back, and writhing in mock death agonies, mouth open and tongue trailing in the dust, it gives a few convulsive twists and wriggles, quiets down, and remains perfectly still and relaxed. It is then supposed to be dead, and if picked up, it will hang as limp as a piece of wet cotton rope and show no sign of life; but if placed right-side up, it will instantly flip over onto its back again and remain still as before. Apparently, from this snake's point of view—if it has any—a dead snake should lie only on its back. If you retreat for a few yards and stay still, it will slowly raise its head, take a cautious look at its surroundings, and if all seems clear, will right itself and crawl away. If you follow the snake after it has started to crawl away and frighten it again, it may repeat the whole performance, just as if it were "putting it over you" with a new stunt.

This, like much animal bluff, is instinctive rather than reasoning behaviour, and is one example of the limitation of instinctive behaviour to established routine, referred to on page 6. Meeting man is a new experience for the hog-nosed snake, and no other enemy was ever intelligent enough to put its bluff to such tests. Yet, it is not clear, as must be admitted, how playing dead would afford it any protection from a snake-eating enemy.

Usefulness of Snakes

ONE OF THE COMMONEST questions I have been asked about snakes is "what use are they?" What is meant is, of course, what use to man, and, strictly practical use. The same question may be asked about a multitude of things in nature, and often no satisfying answer can be given. Even when we cannot find an answer, we have no right to judge, because we know so little.

However, for snakes a practical answer can be given. The most apparent usefulness of snakes lies in the fact that they are carnivorous, that is, they eat other animals. Without the services of carnivores, including snakes, what we call "the balance of nature" and accept as a necessary, self-preserving state of stability, would soon cease to exist.

The reproductive powers of all animals are beyond the ability of their environment to maintain them. In other words, every species would, if left without any checks or controls, eventually produce more young than there was enough food to feed. This is especially true of such fast-breeding animals as rodents (rats, mice, gophers, and their relatives), and many kinds of insects. Since nature does not put controls on the act of reproduction itself, the essential controls which eliminate the surplus

14

offspring have to be applied later (after the young are born). The most valuable of these controls is in the role played by the carnivores; so when the numbers of carnivores are reduced much below the required level, pest animals multiply freely and run riot.

A number of our smaller snakes are insect-eaters, and some are slug-eaters. All our larger ones, except the water snake and hog-nosed snake, feed largely or chiefly upon rodents, of which they destroy vast numbers. Snakes occupy a special place among rodent-eaters, because they ferret out the nests of these animals hidden away in sheltered places, where they are securely protected from hawks, owls, foxes, skunks and other carnivores, and there devour whole litters of young before they are old enough to leave their nests, do any damage, or breed. Snakes also hunt unhindered through dense tangles of brush, thorns and herbage where small rodents find safe refuge from other enemies.

Our debt to these despised and persecuted creatures is never forced upon our attention, for the simple reason that the pest animals that they destroy never annoy us afterwards, neither do the teeming millions of offspring which they would have produced had they lived. We remain blissfully unaware of the fact that they act as a screen between us and the possible damage of swarms of destructive little enemies. When such screen animals as snakes are killed off because of our stupid dislike of them, and outbreaks of pests occur, eating up our crops and stored products, and girdling our fruit trees, we lose millions of dollars; we also spend large sums of money on artificial controls, not always very successful.

Snakes and other carnivores perform this same service for us, day and night unnoticed, without fuss or fanfare, and present a bill for exactly nothing. All they ask of us is that we leave them alone to continue serving us. Let us be fair and reasonable toward them; it is in our own best interests. They cannot serve us after they are dead.

Oh yes, I know what you are thinking at this point: you want to ask, but what about rattlesnakes, are we to spare them too? Though I love and respect snakes, both for the marvellous creatures they are, and the service they render, I would be the first to admit that rattlesnakes cannot be tolerated close to human habitation, because their direct threat to human safety is too great (though it does not compare with some other common dangers that we tolerate every day with little concern). However, they should not be hunted or killed in the wilder tracts of country where they mostly live. People visiting such places on business or pleasure should learn to recognize them, leave them alone, and use such reasonable precautions as may be necessary for their own safety.

Many of the casualties in rattlesnake country result, not from the snakes attacking man, but from man attacking the snakes. Nearly 7 per cent of all recorded snake bites in North America resulted from man actually handling the snakes. In this connection, I might mention that the few deaths from rattlesnake bite in Canada in this century, known to me (two in British Columbia and one in Ontario), resulted either from picking up a snake, or from a clumsy attempt to kill it. In none of these cases was the snake the aggressor. If you see a rattlesnake in wild country, then, leave it alone. It will not chase you. It is a very useful animal, and surely you can think of some other useful things in nature that may also be dangerous in some circumstances.

Popular Beliefs

SOMEBODY ONCE SAID, "a lie can travel a thousand miles while the truth is getting its boots on." He might have added that it often takes a long time for the truth to catch up, and by the time it does, many people are so used to the lie that they are suspicious of the truth. It is like that with snake myths, and perhaps that is why the same old beliefs are passed along from generation to generation to be as firmly rooted as ever.

Many animal myths begin with an incomplete observation, or a coincidence, from which the rest of the story is inferred. As these old stories are told and retold, they lose nothing in the telling, so you may hear different versions of the same myth, some showing "improvement" over others.

Perhaps the commonest of all popular beliefs about snakes is, that a mother snake, when danger threatens her young, will open her mouth and let them crawl down her throat and hide there until the danger has passed, or until she has escaped with them to safety. The Royal Ontario Museum has received many letters from "eye-witnesses" who claimed to have actually seen this happen and has also had telephone calls and visitors affirming the same thing. One interesting fact emerged from all these "eye-witness" stories: old people had seen it when they were young, middle-aged people had seen it when they were children, but nobody had seen it recently—within from twenty to thirty or more years of the time of telling us about it.

Should we then assume that the habit was once very common among snakes, but all of a sudden they all stopped doing it, say, about the year 1920? You will agree that this is not a very good explanation; wild animals do not change their habits as quickly as that. It seems much more likely that the "slip-up" was on the human side and might be due to a long memory of a faulty observation. Time does sometimes play tricks on the memory, and after the passage of years people may "remember" seeing what they had been taught to expect. Certainly, if the habit was as common as the hundreds of "eye-witness" reports by untrained people would suggest, you might reasonably expect it also to have been seen occasionally, or even once, by a trained herpetologist who spends most of his life studying snakes. Nevertheless, no herpetologist has ever seen it, in the field, in a zoo, or in a zoological laboratory. Also, it should be possible to put it to experimental test and occasionally get a positive instead of a negative result.

The belief might have got started from either of two observations, to which a wrong meaning was attached. (1) Some snakes do swallow smaller snakes (not their own young), but the swallowed snakes are soon digested, and that is what would happen to a snake's own young should they go down her throat. (2) Some snakes, as already pointed out on page 8, give birth to active young. If a snake ready to produce active young were killed, and hacked up a bit (as is usually the case), the young that had not been crushed would escape and crawl away. A person witnessing this, and not knowing that some snakes give birth to active young, might suppose that they had first been swallowed for protection.

Almost as common, but with perhaps a smaller battalion of "eye-witnesses" to support it, is the belief that snakes suck milk from cows. In North America some of the king snakes, which are therefore called milk snakes, are the supposed culprits, and other kinds of snakes in other parts of the world. The snakes suspected are species that feed upon rodents, and therefore frequent farm buildings in search of rats and mice. Even if you did not like milk, but were very thirsty and could get nothing else to drink, you would drink it and be thankful to get it. Likewise, a thirsty milk snake, if unable to find water in dry weather, might as conceivably accept milk that was exposed in an open vessel. The milk yield of cows fluctuates from time to time, and may fall off noticeably when pastures dry up in very hot, dry weather. Here we have the stage setting for the milk snake story: the cows were giving less milk, snakes were seen about the barn, perhaps even drinking milk from an open vessel. It is then merely a matter of plain logic and simple arithmetic to add two and two together and make five or six: the snakes were milking the cows.

There are several practical difficulties in this explanation of what happened to the missing milk. The capacity of a full grown milk snake's stomach might be as much as three cubic inches, while there are nearly twenty-nine cubic inches in a pint of milk, so even if the snake filled its stomach with milk every day it would drink less than a pint in a week; if it filled its stomach three times a day, it would still drink only slightly more than a quart. Another of the difficulties is that a snake's mouth is not constructed for sucking milk, and even if it could do so, it contains six rows of needle-sharp teeth. Now use your imagination.

Another widely believed myth is the one about the "hoop snake." This imaginary reptile is said to have a deadly sting at the end of its tail, and to give chase by putting its tail in its mouth and rolling along like a hoop. When it catches up with its victim, it withdraws its tail from its mouth and kills him with one stab. Should the intended victim dodge behind a convenient tree, so that the snake misses him and strikes the tree instead, the tree dies very soon after.

Once, a young man, who was old enough to know better, told me that when he was a boy on the farm, his father was chased by a hoop snake. He related the story of this life-and-death chase in great detail, even to the rapid death of the intervening tree. It had made a strong impression on his boyhood mind. How did he ever come to "remember" this? Did somebody tell him the hoop snake story when he was very young and did he then dream about it?

The story of the hoop snake probably evolved out of an older, common belief that some snakes sting with their tails. This belief is held about the fox snake by some people in Ontario. A fisherman at Long Point on Lake Erie, where this snake is common, informed me that he had seen one sting a potato with its tail, whereupon the potato turned black and began to shrivel. In the southern United States there are two species of snakes, the rainbow snake and the mud snake, whose tails terminate in conical scales, with which they probe about when held, just as if they were trying to stab. Some authorities suggest that the hoop snake story had its origin there.

Equally unfounded in fact is the belief that some snakes have poisonous breath, toxic enough to cause illness, or even death, to man or beast that comes within range of it. In Canada, the harmless hog-nosed snake (page 38) is the one most feared on this account. In farming communities where this snake occurs, you may meet the occasional "eye-witness" who will assure you that he has lost horses or cattle from this cause. The equally harmless fox snake is another species more or less feared for the same reason; in my snake hunting expeditions in southern Ontario I have met "eye-witnesses" who told me of the distressing illness that results to human beings from exposure to its breath.

When I was a child I believed that snakes could, by the mysterious influence of their glassy stare, "charm" their prey, and thus force it to stay still, or even come to them. I believed this because I was told it by older people. Of course, I had not seen it happen, nor had anyone else, but I had seen the glassy stare (page 3); so the rest of the story seemed quite reasonable. When I was older I learned how snakes can sneak up on their unsuspecting prey without being seen or heard. The front view that a snake presents to an animal on ground level is very small, and it can glide along without making any noise. The observable facts of the glassy stare and the stealthy approach may, together, have given rise to the belief, which is widespread and very old. An animal (or man) may sometimes be paralysed by fear at the approach of an enemy. It does not happen very often, but when it does, it has nothing to do with "charming" by any mysterious power in the eyes of the "charmer." That kind of power does not exist in the animal kingdom.

Still another fantastic belief is that a snake will not die until sunset, no matter how severely injured it may be. This belief was probably derived from the observable fact that a snake's body may cling to life for a while after a fatal injury or mutilation and may survive an injury that would quickly kill any higher animal. However, the hour of its death, when it comes, has no connection with the setting of the sun. Probably connected with this belief, is the even more fantastic one, that if a snake is chopped into pieces, the pieces may wriggle together and reunite, so that the snake is again "as good as new."

When I was a young boy I "learned" that rattlesnakes will not cross hair ropes; that they go about in pairs, and will avenge the death of their mates. At about the same time I "learned" that some tropical snakes attain fantastic sizes, hundreds of feet for big pythons and boas, and up to a quarter of a mile for "sea serpents." There is nothing extraordinary in a young child believing such nonsense, were it not for the fact that he must first "learn" it from adults, or from other children who, in turn, learned it from adults.

When I was old enough to read about the natural history of snakes for myself, my ideas about possible size got deflated unbelievably. Nevertheless, when I learned that no giant snake of even as great a length as thirty-five feet had ever been collected and actually measured, it did not spoil my fun. I was just as happy in a world with thirty-foot pythons and no "sea serpents."

Of course, there are no giant snakes in Canada, but that does not mean that there may not be some over-stretched ideas about the length of some of our snakes. No Canadian snake ever exceeds eight feet, a length perhaps rarely attained by the black rat snake (pilot black snake). The fox snake and racer come next, with a maximum length of perhaps six feet. Nevertheless, a farmer at Point Pelee once assured our museum party there that he had killed a sixteen-foot fox snake, and to prove it, he showed us the exact place where he had hung its dead body on the fence. You cannot argue against "proof" like that.

What I have said above does not exhaust the myths and wonder stories about snakes, but we must leave this subject now because of limitations of space. You may have heard some of these ridiculous stories, and even believed them, and you may hear many more before you are old. However, the natural history of snakes loses nothing by "debunking" because many of the facts are more wonderful than the fairy tales.

Fear of Snakes

FEAR OF SNAKES is not natural in children, it is acquired, directly by teaching, or indirectly by example of older people. It is true that in some tropical countries there is no rule by which dangerous snakes may be distinguished from harmless ones, and in such countries caution must be used until one learns to know the different kinds of snakes. This is not so in North America where there are only a few species of poisonous snakes, and simple rules that may be used to distinguish them from harmless ones. In Canada it is still simpler, since the only poisonous snakes are rattlesnakes, and therefore, any Canadian snake with a pointed tail is harmless.

If one of your friends could not learn to distinguish between puppies and kittens, or squirrels and common brown rats, you would probably think he was not very bright. The difference between a Canadian poisonous snake and a harmless one is just as plain and simple and easy to learn. Even if you had never seen a rattlesnake—and many people have not—all you would have to do is to look at a picture of one in any good natural history book or nature magazine and remember what you see. Yet, for all that, many people in Canada are unable to distinguish a harmless snake from a dangerous one.

They are led further astray by the common practice of attaching dangerous-sounding names to harmless snakes, as, for example, "puff adder" or "spreading adder" to the hog-nosed snake, and "spotted adder" or "hardwood rattler" to the milk snake; also by the confusion of harmless native species with venomous exotic species, as the fox snake with the copperhead, or the water snake with the water moccasin, and then reporting them, often in the newspapers, as true records. The newspapers usually accept this nonsense and publish it as fact, without taking the trouble to check with the nearest naturalist or museum. If they do check, it is usually not until the silly story has been published, and many people have believed it because they saw it in print.

As a result of all this, a foolish terror of snakes overshadows the minds of thousands of people when they go to the country for a holiday. A surly and ever-present guest, this fear does much to spoil their own lives, but does not stop there, for by word or example, they pass it on to their children. In this way many children, while still very young, acquire a morbid dread of snakes that haunts them as long as they live. Like all useless fears, it is a burden they can ill afford to carry, and which, like most fear, does them more harm in the end than the thing which they feared.

You may hardly believe this, but mothers frantic with terror have telephoned to me at the museum, because they had seen a few small and harmless snakes in their gardens. They dared not let their children out to play until something was done to get rid of the snakes. They wanted me to tell them what to do. I did tell them—but not what they expected to hear. I told them that small snakes were a part of the southern Canadian scene just as truly as were robins and pine trees. All the wild creatures were here long before the white man came and will continue to be indefinitely. People must learn to accept this fact, share the country with them, and stop fussing about them, or they will make their lives miserable and themselves ill.

Those poor, silly women were like that because older people had made them so while they were still too young to think for themselves, and now they were doing the same cruel thing to their own children, and yet believing they were being kind. I felt as truly sorry for them as I would if they had been brought up to live in constant terror of ghosts and fairies, and I told them so. They usually ended the conversation by saying "Oh, I suppose you are right—but I *hate* snakes!" I could do no more than tell them that the more hates they have, the more unhappy their lives will be, and the only cure for such unhappiness is to think differently.

HARMLESS SNAKES

THERE ARE fourteen species of harmless snakes in Ontario. Two of them, the northern water snake and the eastern garter snake, are each represented by two races (see synoptic list, page 64). The harmless snakes may be recognized as harmless by the following features: the tail is without a rattle, and tapers to a point (Figures 2I, J, p. 71); the eye pupil is round (Figure 2B); there is no pit (Figure 2, A to C) in the side of the face between the eye and the nostril. A harmless snake may acquire a blunt tail as the result of an injury, but it never has a rattle, and you can still see that the eye pupil is round, and that there is no facial pit.

These distinctions would not apply in all parts of the United States, where some poisonous snakes (water moccasins and copperheads) have pointed tails, and some others (coral snakes) have pointed tails, round eye pupils and no facial pit. There are no water moccasins, copperheads or coral snakes in Canada, where the only venomous snakes are rattlesnakes; so the rule outlined above is a perfectly safe one to follow in Ontario.

THE QUEEN SNAKE

SIZE AND STRUCTURE. This snake reaches a length of about thirty inches. It is moderately slender, its head is shallow, and the scales on its body are dull and keeled (see Figure 2E).

COLOUR. It is brown above, with a broad yellow stripe along each side. There are usually three narrow black lines, one down the middle of the back, and one along each side of it, three scales above the yellow stripe (on the fifth row of scales). The belly is yellow, with a dusky stripe along each side of the middle.

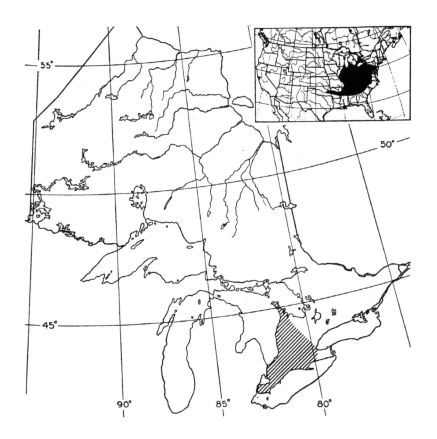

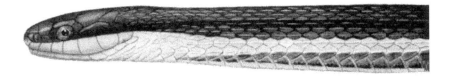

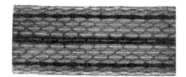 

HABITS AND HABITAT. It is an aquatic snake, and an agile swimmer and diver. It frequents the margins of streams, preferring those with a slowish current, and a stony or gravelly bottom. However, it has been found about still water in old quarries and canals. It has some climbing ability and is fond of resting in low bushes or on tree trunks, commonly willows, overhanging the water, into which it can glide or drop if alarmed. This habit probably gave rise to the name "willow snake" applied to it in some parts of its range. There it dives to the bottom and hides beneath stones or mud. Being alert and shy, it is a rather difficult snake to capture. As one writer (Roger Conant) put it, "it may be rushed successfully but the collector is as apt as not to get nothing but a bath for his trouble." When caught in the hand it usually defends itself by thrashing about and emitting a disagreeable secretion from the scent glands. Some queen snakes will bite, others will not: the conflicting reports by different collectors about their behaviour in this respect suggest that local populations may differ from each other in temperament, as is sometimes true of other snakes. It fares badly in captivity, often dying within a year. It feeds chiefly upon crayfish, preferably when the shells are soft soon after molting, and to some extent upon fish and frogs.

The young are born between early July and early September. They may number from five to eighteen in a litter, but usually not more than twelve.

DISTRIBUTION IN ONTARIO. It has been found in the counties of Brant, Bruce, Huron, Kent, Middlesex and Waterloo. It is not common.

25

THE NORTHERN WATER SNAKE

SIZE AND STRUCTURE. This snake attains a length of four feet, and occasionally a little more. The body is thick and the head deep. The head scales are shiny, the body scales dull and keeled.

COLOUR. In adult specimens the general colour above is brown. There is a mid-dorsal row of squarish dark blotches, and a lateral row of smaller ones on each side. The lateral blotches alternate with the dorsal ones on most of the body length, but on the forward quarter (more or less) they line up with the dorsal blotches, forming cross bands. These markings become duller with age, and may blend into a uniform brown in old snakes. The belly may be creamy, yellowish or reddish, commonly with dusky mottling. Newly born specimens have nearly black blotches and bands on a whitish ground colour.

HABITS AND HABITAT. This is a very aquatic snake, frequenting lakes, streams, ponds and marshes. It is an excellent swimmer and diver, very much at home in the water, from which it never wanders far. It is fond of basking on the shore or on stumps or rocks in the water, but is shy and watchful and difficult to approach. It will bite viciously when first caught, so many collectors wear leather gloves when catching it.

It feeds chiefly upon fish, but frogs, salamanders and crayfish are also eaten, and occasionally small snakes, mice and shrews. Examination of

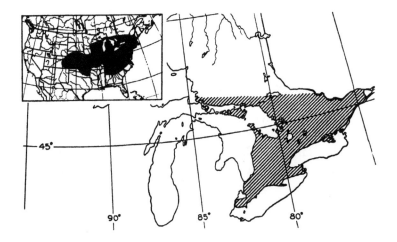

the stomachs of many water snakes has shown that this snake, except when living at hatchery ponds, takes very few game fish; most of its food in natural waters consists of other kinds of fish, some of which, such as sculpins (millers' thumbs), are enemies of game fish. By eating dead and diseased fish, it helps to keep natural waters clean and healthy. It should be protected and not killed. It is a useful species.

The young are born in August or September, and usually number from twenty to forty in a litter.

DISTRIBUTION IN ONTARIO. The water snake is found in the southern portion of the province, and northward into the southern parts of Algoma and Sudbury districts. It is generally common south of Parry Sound where suitable situations exist.

A pale-coloured, nearly uniform gray race of this snake, called the Island Water Snake, is found on Pelee Island, and some other islands in western Lake Erie.

DEKAY'S SNAKE

(Little Brown Snake)

SIZE AND STRUCTURE. This small snake attains a length of about fifteen inches,
rarely more. The body is rather stout; the dorsal scales are dull and
keeled.

COLOUR. The colour above varies from light to darkish brown, or grayish
brown. There is a lightish vertebral stripe equal in width to four scales,
bordered on each side by a slightly darker tone, and by a row of dark
spots which often encroach upon it. A second, lower row of alternating
dark spots is sometimes visible. The belly is pale brown or oatmeal-
coloured, or slightly pinkish.

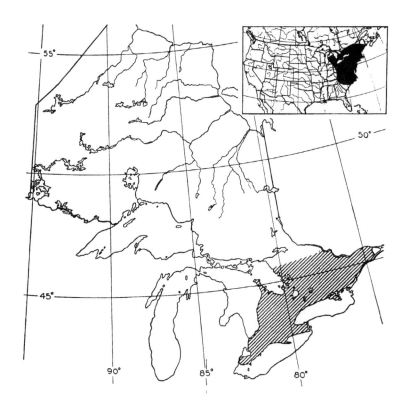

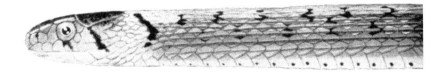

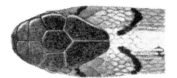
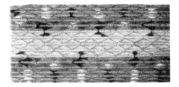

The newly born young are nearly black above, with a light ring around the neck. Their keeled scales will serve to distinguish them from young ring-necked snakes.

HABITS AND HABITAT. The little brown snake inhabits light woods, clearings, farms, fields and roadsides; it often lives in parks, gardens and vacant lots in built-up areas. It is secretive and, except in the spring and early autumn, remains in hiding during most of the day, coming out in late afternoon. Its food consists mainly of slugs, snails and earthworms; indeed, slugs and snails form over 60 per cent of its diet. It is therefore a very useful little snake about farms and gardens. No doubt its small size, dull colouration, secretive ways, and the ease with which its food may be found, all combine to help it to survive in city parks and vacant lots long after other species have disappeared from them.

It requires some dampness in its environment and generally avoids completely dry places. This gentle little snake, which never attempts to bite, makes an attractive pet, becoming very tame and doing well in captivity, if its simple requirements of food, water and slightly dampish cover are met.

The young are born from late July to early September and may number from about nine to twenty in a litter.

DISTRIBUTION IN ONTARIO. DeKay's snake is found throughout all of southern Ontario, northward in Parry Sound District to Pointe au Baril, and perhaps to the French River. It is common.

THE RED-BELLIED SNAKE

SIZE AND STRUCTURE. Our smallest snake, rarely exceeding one foot in length. The body is moderately stout; the body scales are dull and keeled.

COLOUR. The colour above varies from light to dark brown or gray, chestnut, or rarely black. There is a lightish vertebral stripe three scales wide, bordered on each side by a dusky line; another dusky line on the first row of scales of each side. A yellow spot is present on the back of the neck, and one on each side. The belly is red or pink, with a band of dark speckling along each side. (See Plate II A.)

The newly born young are very dark, with a whitish neck ring.

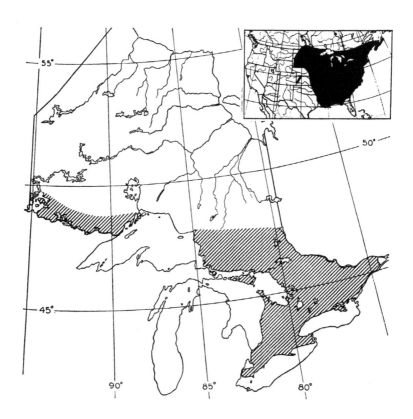

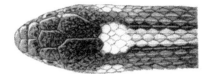 

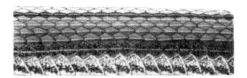

HABITS AND HABITAT. The red-bellied snake is very similar to the little brown snake in its ways of life. Most of the specimens which I have collected were found beneath stones, boards, or other cover on the ground. Like the little brown snake, it remains concealed during most of the day, coming out of hiding and wandering about toward evening. It may, however, be found abroad in the daytime in spring and early autumn. I have often found it basking in warm afternoons in October near Toronto. Although its habits and general requirements are, in general, so similar to those of DeKay's snake, it is, apparently, a hardier species. Indeed, next to the common garter snake, it has the widest and most northerly extended range of any Canadian snake.

It feeds upon slugs, earthworms and beetle larvae. Slugs are apparently its main food; it is thus a very useful snake to the gardener and should be encouraged and protected.

Like DeKay's snake, it requires some moisture in its surroundings. It is entirely gentle, makes a pleasing pet, and will live very well in captivity if its requirements in food, water, and slightly dampish cover are met.

The young are born in August or September and may number from one to thirteen in a litter, but usually seven or eight.

DISTRIBUTION IN ONTARIO. This little snake is found all over southern and central Ontario, northward to New Liskeard, Gogama and Quetico Park. It is common.

BUTLER'S GARTER SNAKE

SIZE AND STRUCTURE. This is a smallish garter snake, attaining a length of about twenty-two inches. The head is small and short, scarcely wider than the neck. The eye is smaller, and the body relatively a little stouter, than in the eastern garter snake. The body scales are keeled.

COLOUR. The colour above is brown or olive brown, with three yellowish stripes, a mid-dorsal and two lateral. *The lateral stripe of each side is centred on the third row of scales on the forward quarter of the body,* but lies on the second and third rows posteriorly. The first scale row and the sides of the belly are chestnut. There is a lightish spot in front of the eye.

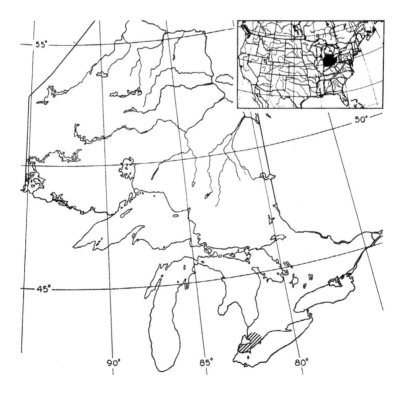

HABITS AND HABITAT. Butler's garter snake is usually found in the vicinity of water, near marshes, along the shores of lakes and streams, and in wet meadows. However, in middle and late summer it may be found in dry places that had been wet in the spring. There is a colony of this snake in Middlesex County in such a place. We found the snakes in dry grass in a clearing and along the side of an adjacent road. Although the immediate situation was then dry, there was a ditch with some water just across the road. At the time, in early July, the weather was very hot and the snakes were active only in the evening from sunset until dusk.

When removed from the grass and placed on the dirt road, they were clumsy and ineffectual in their attempts to hurry, but in the grass they glided smoothly and swiftly. When picked up in the hand Butler's garter snake behaves differently from the eastern garter snake. It partially wraps itself about the fingers and clings with perceptible pressure, actually using its tail as a prehensile (grasping) organ.

The species feeds upon earthworms, fish and frogs. The young are born in July or August; litters numbering from four to nineteen have been counted.

DISTRIBUTION IN ONTARIO. This snake is found in extreme southern Ontario. Actual records are from Kent and Middlesex counties only, where it seems to occur locally in colonies. It should be looked for in all the western Lake Erie, and the Lake Huron, counties.

THE EASTERN RIBBON SNAKE

SIZE AND STRUCTURE. This is a slender snake with a rather lizard-like appearance about its head, and a long tail, noticeably longer than in our other garter snakes. It reaches a length of about thirty inches. The eye is large. The body scales are keeled.

COLOUR. The dorsal colour is olive to dark brown or black with three brilliant yellow stripes, a vertebral and two lateral. *The lateral stripe of each side lies on the third and fourth scale rows,* and is usually bordered below on the first scale row by a chocolate brown band. The chin, throat and upper lip plates are bright yellow. There is a vertical bright yellow spot in front of the eye. The belly is greenish.

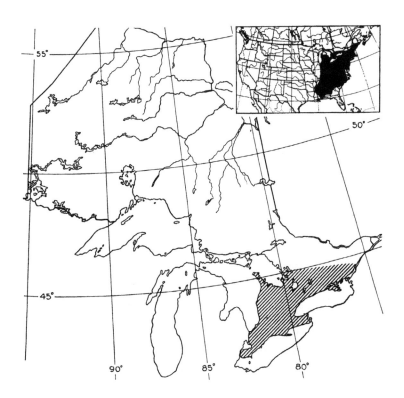

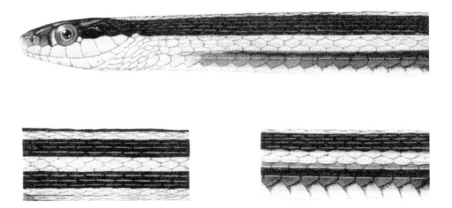

The sharp, clean-cut appearance of the yellow stripes, along with the very slender body, are good field characters for quick identification.

HABITS AND HABITAT. The ribbon snake inhabits damp situations about the margins of streams and ponds, the low, grassy and weedy parts of the shores of lakes, wettish meadows, springs and bogs. It is moderately aquatic and a good swimmer, readily entering the water if pursued. On land it is swift and graceful, appearing to glide without effort. It is also a fair climber, occasionally found at several feet above ground among the lower limbs of bushes.

The ribbon snakes which I have caught and handled were nervous, but mild-tempered, struggling wildly at first and using their scent glands, but rarely (if ever) attempting to bite. In captivity some remained nervous, others settled down and learned to take food from the hand. The less nervous specimens make attractive pets.

The food of the ribbon snake consists mainly of amphibians and small fish. Insects may be eaten, but it seems likely that many of those found in their stomachs had first been eaten by amphibians which the snakes swallowed afterwards. Although it has been reported as eating earthworms, none of our specimens would ever take them.

The young are usually born in August; they may number from three to twenty, but more commonly from five to twelve, in a litter.

DISTRIBUTION IN ONTARIO. It is found in the southern counties only, northward into Muskoka District. It is moderately common in some localities where suitable conditions exist.

THE EASTERN GARTER SNAKE

SIZE AND STRUCTURE. The eastern garter snake is a moderately slender-bodied species, which attains a length of about forty inches, rarely more. Most adult examples range between twenty and thirty inches in length. The tail is noticeably shorter than in the ribbon snake. The eye is large, but smaller than in the ribbon snake. The dorsal scales are keeled.

COLOUR. The dorsal colour may be black, olive or brown, normally with three yellowish stripes, a dorsal and two lateral. *The lateral stripe of each side lies on the second and third scale rows.* Usually there is no bright spot in front of the eye. The belly is yellowish or greenish. At Long Point and Point Pelee on Lake Erie, black specimens with a white throat and chin are not uncommon. The species is highly variable in its colouring, and in the brightness or dullness of its striping.

HABITS AND HABITAT. This is the commonest and most widely distributed snake in Ontario. It ranges further north than any other, is the first to appear in the spring and the last to disappear in the autumn. There

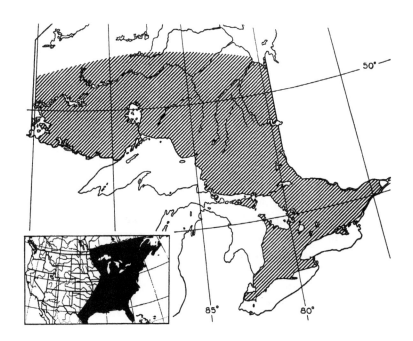

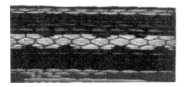

are occasional records of garter snakes coming out on very mild, sunny days in winter. Winter basking is a risky pastime for snakes, however hardy they may be, especially so if they dare to wander even a short distance from their dens. The dead bodies of these snakes are occasionally found frozen on top of the snow.

The species frequents a great variety of country, being found in woods, clearings, on farms, along roadsides, in marshes, on the shores of lakes, ponds and streams. It prefers to be near water, but also wanders far from it into high and dry places.

It feeds chiefly upon earthworms and amphibians, and small fish when available. It occasionally eats other snakes, small birds and mice. The prey is usually caught and swallowed alive, but dead animals are readily accepted. At the museum we fed our captive specimens mostly upon pieces of raw fish, dead frogs, and earthworms.

Most individuals are good-tempered, but some will strike repeatedly when caught; they all use their scent glands at first. This snake adapts itself well to captivity and makes a good pet.

The young are born from July to October and usually number from ten to thirty in a litter, but as many as seventy-eight were noted on one occasion.

DISTRIBUTION IN ONTARIO. The eastern garter snake is found all over the province to as far north as James Bay. It is very common.

It intergrades with a plains race, the red-barred garter snake, in extreme western Ontario.

THE EASTERN HOG-NOSED SNAKE

THIS HARMLESS SNAKE is also called by such dangerous-sounding names as Puff Adder, Blowing Adder, Spreading Adder and Sand Viper.

SIZE AND STRUCTURE. The hog-nosed snake is very stout-bodied and reaches a length of about three feet. The head is short and broad; the snout pointed and protruding, slightly upturned at the tip. The tail is very short. The body scales are keeled.

COLOUR. This snake is variable in colour, but most commonly olive, brown or gray above with a pattern of darker blotching. There is usually a mid-dorsal row of large darkish blotches, and a row of smaller ones along each side alternating with them. Sometimes the body blotches are faint or absent, but there is always a pair of large, elongate ones, one on each side of the neck. In very dark or black specimens the blotching is obscured.

HABITS AND HABITAT. The hog-nosed snake prefers sandy situations, beaches and light, dry woods. It is an entirely inoffensive snake, but its extraordinary behaviour when frightened, as described on pages 13, 14, has given rise to fantastic beliefs about its supposedly dangerous nature. The superstition about its poisonous breath is described on pages 18–19.

The neck-spreading practised by this snake is accomplished in the same manner as it is by a cobra when it spreads its hood, that is, by especially elongated movable ribs that lie folded backward when at rest.

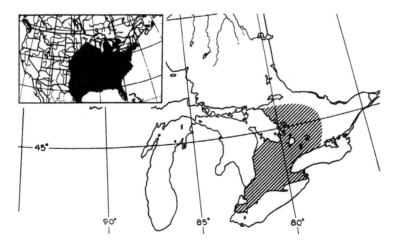

The hog-nosed snake feeds chiefly upon toads, and to some extent upon frogs. Some other kinds of animals have rarely been reported as eaten by it. Certain of its structural features appear to be related to its habit of eating toads. In the rear of the upper jaw is a pair of long, fang-like teeth, but without grooves or poison glands, for the snake has none. The usefulness of these teeth probably lies not only in their ability to hold the large and struggling prey, but also in puncturing and deflating its body, because toads inflate themselves with air when attacked by snakes.

The eggs, from about twelve to possibly thirty, are laid in dampish places, such as the pulpy wood of decaying logs, in late June or July.

DISTRIBUTION IN ONTARIO. The hog-nosed snake is found in southern Ontario northward to northern Parry Sound District and southern Algonquin Park, eastward to Durham County, and possibly into Hastings County. It is moderately common in some areas.

THE EASTERN RING-NECKED SNAKE

SIZE AND STRUCTURE. A slender snake, occasionally reaching a length of about eighteen inches; most specimens are smaller. The head is rather broad and flattened. The scales are glossy and unkeeled.

COLOUR. Gray to bluish slate above. The belly is yellow or orange, and there is a ring of the same colour around the neck. There is usually a row of dark spots along each side of the belly. (See Plate II B.)

HABITS AND HABITAT. This very secretive snake inhabits woods and clearings, and stony pastures near woods, if not too dry. In the day time it is

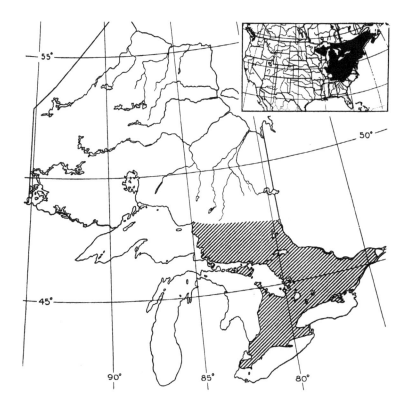

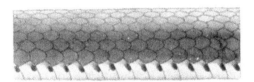

usually found hiding beneath stones or in stone piles, or under the bark of decaying logs or stumps. It is apparently nocturnal in its wanderings. I have never seen it abroad before twilight, and specimens uncovered in the full light of day seem more concerned to avoid the light than to glide away.

It feeds chiefly upon small snakes, lizards, salamanders and frogs, but insects and earthworms also enter its diet. Most of our captive specimens, however, refused earthworms, but accepted red-backed salamanders and young snakes.

It lays its eggs in dampish places, usually in the soft wood of rotting logs, and just beneath the firm, outer shell of non-pulpy wood. For this, July is the preferred month. The individual clutches are small, mostly three or four eggs, rarely up to seven. This snake has a tendency to use community nests in which the clutches of several females are laid together.*

DISTRIBUTION IN ONTARIO. The ring-necked snake is found in the southern part of the province and northward into southern Nipissing District and southern Algoma District. It is not common over most of its range in Ontario and tends to occur in local colonies.

*Dr. Blanchard reported one nest (in Michigan) that contained 48 eggs, to which about 14 females must have contributed. That the same nests may be used in successive seasons was apparent by the presence in some of them of empty egg shells of previous summers.

THE EASTERN BLUE RACER

SIZE AND STRUCTURE. The blue racer is a large, slender snake which occasionally attains a length of six feet. The sides of the face are high, with a
concavity (hollow) in front of each eye. The crown of the head is flat,
the profile then sloping downward to the snout. The eyes are large. The
dorsal scales are satiny and unkeeled.

COLOUR. The colour above is a uniform bluish green; the under surface is
bluish or greenish white, or yellowish.

The young are coloured quite differently: they are pale gray, with a
row of large, dark dorsal blotches, and a row of smaller alternating ones
along each side. They resemble milk snakes, with which they may be
easily confused, but may be distinguished from them by their very large
eyes (those of a milk snake being small), and by the lack of any dark
markings on the belly.

HABITS AND HABITAT. The blue racer is reported to prefer rather dry, open
places, and to frequent the vicinity of thickets and brush, loose rock or
old stone structures; it also inhabits woodlands, meadows and the
margins of marshes. It is rare in Ontario; so we have not much local

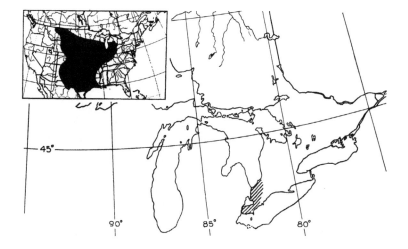

information about its habitat preferences. Fishermen at Point Pelee on Lake Erie informed me that this snake was sometimes seen sunning itself at the edge of the marsh in spring before the vegetation had grown tall, but later in the summer it was seldom seen. They could never get very near to it before it would dash away.

It is very alert, and swift for a snake, and although mainly a ground snake, it is also a good climber. It is reported to be rather local in its habits, having particular basking places and hiding places to which it returns again and again. When first caught it is nervous and pugnacious, striking and biting freely. Some specimens in captivity may remain wild and hostile indefinitely; others settle down and become tame pets.

The racer eats mammals, birds and their eggs, reptiles, amphibians and insects. Insects and rodents form the largest part of its food; among the snakes eaten, venomous species are included.

The eggs may number from eight to twenty-five in a clutch, but usually around a dozen. They are laid from late June through July in dampish situations, such as decaying logs, sawdust piles, rubbish heaps, or in loose earth.

DISTRIBUTION IN ONTARIO. Our positive records are from Essex and Huron counties only, but it possibly occurs sparingly over the southwestern tongue of peninsular Ontario from about Norfolk County to Essex County and northward perhaps to southern Bruce County.

43

THE EASTERN SMOOTH GREEN SNAKE

SIZE AND STRUCTURE. This slender and beautiful little snake attains a length of about from sixteen to twenty inches. The body is of nearly the same diameter for most of its length. The scales are of a satiny texture, and without keels. From this feature it receives its common name "smooth" green snake, to distinguish it from the rough, or "keeled" green snake of more southern distribution in the United States.

COLOUR. This snake is a uniform grass green above, pale yellow or whitish beneath (Plate II C). The newly born young are dark gray above and dull whitish beneath.

HABITS AND HABITAT. The smooth green snake inhabits grassy fields, clearings and open woods. Though normally a ground snake, it is a fair climber and may sometimes be found above ground in the lower limbs of shrubs and vines. Its pure green colour above makes it difficult to see when it is in the grass, or among the green leaves of bushes, even when it is gliding smoothly along. It is often found beneath such cover as flat

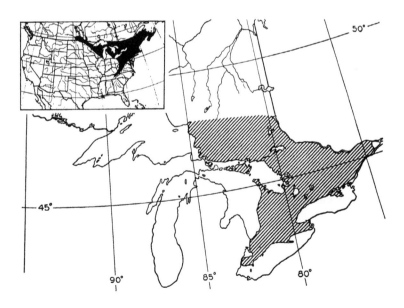

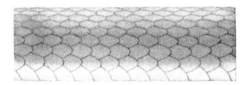

stones, pieces of wood or bark, old sacking, sheet metal or cardboard on the ground. When such cover is lifted, the snake can easily be seen on the brown earth with which its colour contrasts, but this is only for a few seconds; once it darts off into the grass or other vegetation it is nearly impossible to find it again. When I was a boy, the first green snake I found escaped me in that way; I searched the surrounding area long and carefully, but to no avail. That loss worried me for months, and it was several years before I found my next one.

This little snake is very inoffensive, and I have never known one to attempt to defend itself in any way when caught in the hand. It feeds chiefly upon grasshoppers, crickets, caterpillars and spiders; other insects, snails, small salamanders, and possibly small snakes, are eaten occasionally.

The eggs are laid from late July to late August, under such cover as flat stones or boards on dampish ground, and usually number six or seven in a set, although as few as three and as many as eleven have been reported. The eggs of this species are unusual in being well advanced in development at the time of laying, so that hatching is commonly completed within one or two weeks. The longest and shortest hatching periods recorded are twenty-four and two days, respectively. The eggs of most Ontario snakes require from seven to nine weeks.

DISTRIBUTION IN ONTARIO. This snake is found all over southern Ontario, northward to Gogama in Sudbury District. It is fairly common.

THE EASTERN FOX SNAKE

SIZE AND STRUCTURE. The fox snake* is a large, moderately slender-bodied snake, which occasionally reaches a length of six feet. The dorsal scales of the first three to five rows are smooth, those above them on the back are feebly keeled.

COLOUR. The ground colour above is brownish yellow (actually, the individual scales are yellow with brown centres). There is a dorsal row of large, dark brown blotches, and two series of smaller alternating ones along each side. The belly is yellow, with alternating squarish, black blotches. The young are gray, but show the same pattern of blotching.

HABITS AND HABITAT. This snake is to be found only near water, about lake shores, beaches and marshes. Though not really an aquatic snake, it readily enters the water and swims for relatively long distances, sometimes for hundreds of yards across bays, or between islands. It may be seen basking on mats of dead reeds along the edges of marshes, on muskrat houses, and on rocks or wharves at the shores of lakes, when the sun is not too hot. In the early forenoon, and after sundown, it may be found beneath cover on the ground, or in the interior of hollow logs.

The fox snake is usually mild-tempered, but occasional specimens will bite when first handled. Wild specimens commonly use the vile-

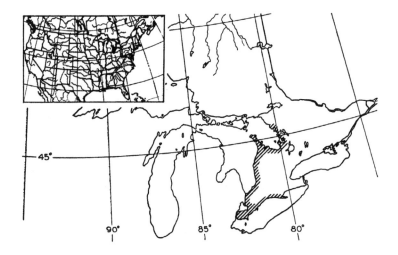

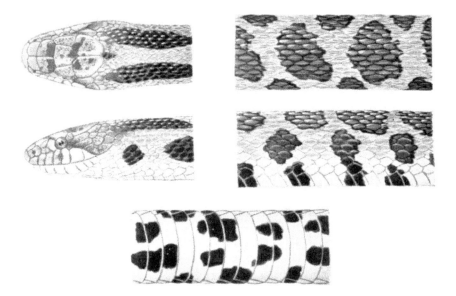

smelling secretion of their scent glands when picked up in the hand. They soon get over their fear and accept gentle handling without apparent resentment, and make quiet and attractive pets.

The species feeds chiefly upon mammals, and birds and their eggs. Rodents are the most important food item. However, other kinds of animals are eaten occasionally: a young specimen caught at Long Point disgorged a bundle of earthworms; and one caught at Go Home Bay vomited up a salamander. Large prey is killed before being swallowed by being constricted in a few coils of the muscular body. This is a very valuable snake.

The eggs are laid in July in damp places such as the interior of decaying logs, sawdust piles or manure heaps, and may range from seven to seventeen in a set.

DISTRIBUTION IN ONTARIO. This snake is found along the shores of lakes Erie, St. Clair and Huron, and at Georgian Bay to as far north as Pointe au Baril. It is common in parts of the region outlined above.

*Because of its habit, shared by various other harmless snakes, of vibrating its tail when frightened or annoyed, it is sometimes called "hardwood rattler," and believed to be venomous. It is commonly confused with the copperhead, a poisonous snake not found in Canada, and most Ontario reports of copperheads are referable to this harmless snake. It is also confused with both the massasauga and the timber rattlesnake.

BLACK RAT SNAKE

(Pilot Black Snake)

SIZE AND STRUCTURE. This is our largest snake, reported to occasionally reach a length of eight feet. Like the fox snake, it is moderately slender, but the belly is much more sharply angular at the sides. The scales of the first three to five rows are smooth, those above them are feebly keeled.

COLOUR. Uniform black above, or marked with obscure black blotches which are large in the dorsal row, but smaller and alternating in the lateral rows. The throat and chin are white. The belly is white or yellowish on the forward part, but becomes increasingly darkened with blotching (often squarish), and other dark mottling, posteriorly.

The young are light-coloured, with conspicuous dark dorsal blotching.

HABITS AND HABITAT. The black rat snake prefers open woodlands, rocky, scrubby, rough country, and uplands, often away from water, giving

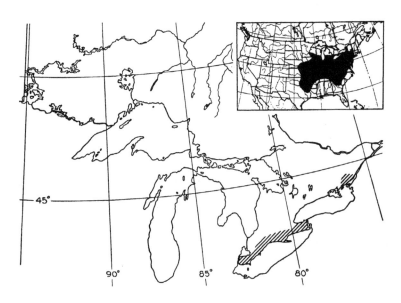

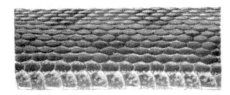

rise to the name "highland" black snake by which it is often known. Like its close cousin, the fox snake, it feeds chiefly upon mammals and birds, but frogs, snakes and other animals are also eaten. The food tends to vary with the season, as most of the birds eaten are taken in early summer, while mammals—largely mice—are the main food in later summer. Like the fox snake, it kills large prey by constriction. It is an economically valuable snake.

When first caught it is sometimes irritable, biting and vibrating its tail, and using the unpleasant secretion of its scent glands. Specimens that we have had in captivity settled down and became gentle and amenable to handling.

The eggs may number from six to twenty-four in a clutch, commonly from twelve to sixteen. They may be deposited in loose earth, sawdust heaps, manure piles, or within decaying logs, from late June through July. The female has been observed to brood the eggs while they hatch.

DISTRIBUTION IN ONTARIO. Records are from the southwestern counties of Essex, Norfolk and Welland, and from the eastern counties of Frontenac and Leeds. I would expect to find it in all the Lake Erie counties, and the Lake Huron counties northward at least to southern Bruce County. Its range appears to be discontinuous north of Lake Ontario. Verbal reports of a large black snake in Prince Edward County, if valid, probably refer to this species. It is fairly common in Frontenac and Leeds counties.

THE EASTERN MILK SNAKE

SIZE AND STRUCTURE. The milk snake may reach a length of three and one-half feet, but most specimens seen are smaller, from two to three feet. The body is slender, and of nearly the same diameter for its whole length. The head is short, broad and shallow. The dorsal scales are glossy and unkeeled.

COLOUR. The ground colour above is pale to medium gray or brown. There are five rows of black-bordered brown blotches: those of the dorsal row are large and saddle-shaped; those of the lateral rows are smaller and alternate with them, and with each other. The belly may be white, but is usually marked with squarish black blotches, or darkened with dusky mottling. In young specimens the dorsal blotches may be reddish.

HABITS AND HABITAT. This beautiful and very useful snake inhabits light woods, clearings, farmlands and rural gardens. It is often found about barns and outbuildings, and occasionally in country houses, where it comes in search of mice and young rats. Rodents form most of its food, in fact, over seventy per cent of it. It also eats birds, lizards and snakes. Since it belongs to the king snake genus, a group that freely preys upon vipers and possesses a fairly high degree of immunity against their

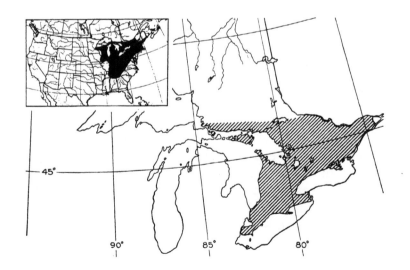

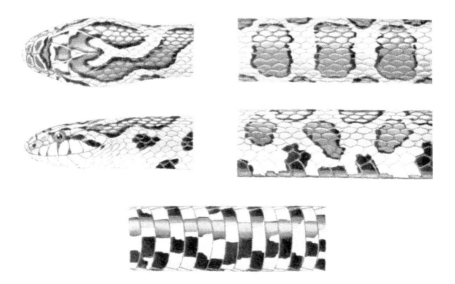

venom, it is probable that young rattlesnakes, when found, are eaten as readily as other kinds. It is a constrictor, killing large prey in coils of its body before swallowing it.

Balancing one thing against another, the milk snake may be considered to be our most beneficial snake. I have met the occasional farmer (unfortunately, they are few) who was wise enough to protect this snake on his property and not to allow it to be killed.

The superstition that it milks cows is discussed under "popular beliefs" on pages 17, 18.

When first caught, the milk snake may be irritable and turn and bite slowly and deliberately. It has a habit of vibrating the tip of its tail when annoyed, thus earning for itself (like the fox snake) the name "hardwood rattler." Many people believe it to be a sort of rattlesnake and venomous. Very young specimens, when caught or cornered, will often hiss and strike repeatedly, but older snakes are less excitable. It often fares badly in captivity, refusing to feed.

The eggs are laid in June or July in rotting logs, sawdust heaps or manure piles, and may number from eight to sixteen.

DISTRIBUTION IN ONTARIO. The milk snake is found in all of southern Ontario, northward to the southern parts of Algoma and Sudbury districts. It is moderately common.

VENOMOUS SNAKES

RATTLESNAKES

IDENTIFICATION. The only dangerous snakes in Canada are the rattlesnakes. There are four kinds, two in Ontario and two in the west. A rattlesnake may be recognized by the following features: the rattle at the end of its tail; its rather triangular head and relatively thin neck; a deep pocket, the pit, between the eye and the nostril; and its cat-like eye pupil which is vertically slit-shaped—not round as in harmless snakes.

COLOUR PATTERN. Rattlesnakes, unless actually black, show some pattern of darker blotches or crossbands on a lighter ground colour. Therefore, when in rattlesnake country, never touch a blotched or banded snake without first taking a good look at the head or tail, preferably at both. To know what a rattlesnake looks like is the first defence against being bitten.

THE PIT. Rattlesnakes belong to the family of snakes known as Pit Vipers. The pit (Figure 3B, D), from which they derive their name, is a pocket-like structure, paired, and situated one on each side of the face between the eye and the nostril. These pits are heat receptors, sensitive to the relative warmth or coolness of any object coming near the snake. Since there is one on each side of the face, they enable the snake to know the direction from which the warmth or coolness is coming, just as human ears, being paired, enable one to tell the direction from which a sound is coming. Since pit vipers are largely nocturnal and often hunt in the dark, the pits are extremely useful: they enable these snakes to detect and strike their prey, even in total darkness, where they could not possibly see it.

THE RATTLE. This structure, which terminates the tail, is composed of a series of loosely interlocking segments of horny material (Figure 3E, F), which, when the tail is vibrated, produce a buzzing sound. The baby rattlesnake has a rounded, horny button at the tip of its tail (see p. 59); one additional segment is added in front of this at each shedding of the skin. It follows then that a rattlesnake never has a pointed tail, so, as already stated (pages 20–21), any Canadian snake with a pointed tail is harmless. As long as the snake is growing, each new segment added is a little larger than the one added previously (the one next behind it on the tail), but when growth stops, the new segments are all of the same size. A complete rattle, then, tapers at the end to the original button, and from it you can tell how many times the snake has shed its skin, but not its age in years. When the segments are all of the same size, which is usual in older snakes, it means that those toward the end of the rattle had been worn or broken off.

THE VENOM APPARATUS. This is by far the most important feature of rattlesnakes. It consists of a pair of long, curved, hollow fangs in the front of the upper jaw, a pair of venom glands, and their ducts which convey the venom to the fangs (Figure 3D). The fangs are movable, and when at rest are folded back against the roof of the mouth and covered by a pair of fleshy sheaths, but when biting or striking, are erected to a perpendicular position in relation to the upper jaw. The fangs are constructed like hypodermic needles, with the lower end of the internal canal opening near the tip, while the upper end is openly connected beneath the sheath with the venom duct from the gland in the same side of the head.

The venom glands are situated one in each side of the face, behind the eye. They are under the voluntary control of the snake, so that it can regulate the amount of venom discharged at a time. Snakes conserve their venom and do not use it all in one bite; so two or more bites from the same snake (which sometimes happens) are more dangerous than one.

STRIKING AND BITING. The typical striking attitude is usually assumed by a rattlesnake when disturbed and on the defensive. The posterior part of the body is then coiled on the ground and the forward part raised and thrown into an S-shaped curve, and the rattle, often raised, may be used vigorously. From this position the head can be lunged forward with great speed for about half the length of the snake; or if the tail is braced against some rigid object, the snake may strike for its full length, but with less accurate aim. Near the end of the stroke the mouth is opened widely, and the erected fangs, which are then directed forward, are imbedded in the victim, discharging their venom deeply into the bite. But a rattlesnake can bite without striking in this manner, as when picked up or stepped on, and the bite so delivered is just as serious.

Rattlesnakes strike either to kill their prey, or in self-defence if they suppose themselves to be threatened, but they are nervous creatures that regard any large animal coming near them as a threat; so give them plenty of space. They will not follow you.

THE VENOM AND ITS ACTION. Snake venoms fall broadly into two natural groups, nerve poisons and blood poisons. Rattlesnake venoms belong to the latter group, but also contain some nerve poisons which attack the nerves controlling breathing and heart-beat. The blood poisons destroy the red blood cells and dissolve the walls of the small blood vessels, and tend to digest any internal tissues with which they come in contact. After rattlesnake bite, therefore, there may be much damage to various internal organs, and haemorrhage (severe bleeding) in the regions so attacked. Death, if it follows, may result from general blood and tissue destruction, or from failure of the nerves controlling breathing or heart-beat, or from a combination of these causes.

Since the primary purpose of a snake's venom is to kill its prey as quickly as possible, it is in some degree selective towards the kinds of animals upon which the snake producing it normally feeds, killing these most quickly. For instance, the coral snakes of the southern United States feed largely upon snakes, and their venom is quickly fatal to such prey, but while as surely fatal to, is slower acting upon, mammals. Rattlesnakes, on the other hand, feed chiefly upon warm-blooded animals, mostly mammals, and their venom is rapidly fatal to these, but much less toxic to snakes. Snakes are tolerant of relatively large doses of rattlesnake venom and often recover from bites that would quickly kill mammals. However, they are not totally immune, and there are many records of their death following a bite, of their own, or another, species.

A mammal struck by a rattlesnake will try to escape if possible, but usually drops within a few feet or yards of where it was struck; if the venom acted as slowly upon it as upon a snake, it would probably get too far away for the rattlesnake to find it. I have seen a mouse die within three or four seconds after being struck by a rattlesnake.

REMOVAL OF FANGS. Many people believe that if a rattlesnake's fangs are removed, it is thereafter harmless. This is dangerous nonsense unless the venom glands are also removed. Rattlesnake fangs (like all snake teeth) are replaced at fairly short intervals—about six to ten weeks—so there is always a reserve supply of developing fangs. Even if the maxillary bone is removed so that no new fangs can appear, but the venom glands are left, venom may still be poured into the mouth, and bites involving only the smaller teeth might be serious.

The fangs and glands are often removed from carnival snakes to make them safe for "snake charmers" to handle for the public. Snakes mutilated in that cruel way die after a few weeks or months; so rattle-snake "charming" is extravagant of rattlesnakes.

THE EASTERN MASSASAUGA

(Swamp Rattlesnake, Little Gray Rattlesnake)

SIZE AND STRUCTURE. This is a thick-bodied snake, which attains a length of from two and one half to three feet. The head is broad, and though not so markedly triangular as in the larger rattlesnakes, is very noticeably wider than the neck. There are nine large, symmetrical scales on top of the head (as in harmless snakes). There is a pit between the eye and nostril on each side of the face. The eye pupil is a cat-like, vertical slit. The tail ends in a rattle.

COLOUR. Gray or brown above, rarely black. A pattern of large, dark blotches down the middle of the back, and three rows of smaller alternating ones along each side. The tail is banded. The belly is mostly black, broken with white mottling toward the neck, the colour changing to gray and white mottling on the throat and chin.

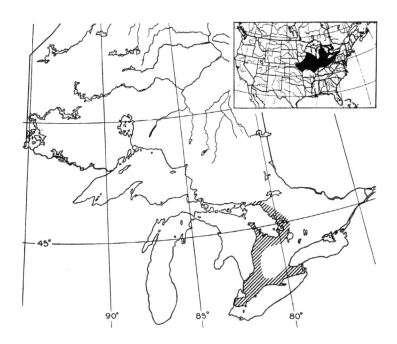

HABITS AND HABITAT. This little rattlesnake lives mostly on low ground and about swamps, but may be found also in higher, dryer situations and rocky places. Though not an aquatic snake, it occasionally enters the water and swims.

It feeds chiefly upon frogs and mice, the latter furnishing its main food supply in middle and late summer, when it often wanders up into hay and grain fields to catch them. Snakes are occasionally eaten. The massasauga is not averse to food that has been dead for many hours and is actually beginning to decay.

It is a mild-tempered snake, and specimens which I have caught in the wild by teasing them into a net or cotton bag with a stick, rattled, but did not strike. Freshly caught wild specimens sent to the museum rarely rattled and never struck. However, I always kept my hands well out of their reach. The warmth of the hand might have caused them to

THE EASTERN MASSASAUGA

strike, because as explained on page 52, the temperature of an object may be a natural stimulus for this act.

The careless handling of this snake is an exceedingly dangerous practice, since it is fully capable of delivering a fatal bite, even though it usually fails to do so. Even if the bite did not prove to be fatal, the severe pain and acute illness that follow should be a sufficient deterrent.

Although the venom of the massasauga is about five times as deadly as that of the Texas Rattler—the chief killer in the United States—the quantity produced by the massasauga is very small, and the shortness of its fangs often prevents it being properly delivered. These facts, along with the mild temper of the snake, must account for the extreme rarity of fatal bites; the more excited and cross a rattlesnake is at the time of biting, the more venom it will deliver.

Although there is some element of risk wherever rattlesnakes are found, the chance of any particular person receiving a bite is extremely small, unless he tampers with a snake, and in Ontario the odds against it are millions to one. The fact that there are a few of these snakes in some of the summer resort areas of southern Ontario need not spoil your holiday. They are not nearly as dangerous as motor cars. Indeed, one takes a far greater risk with his life when driving for a few miles on any highway than he would take with it in a whole summer in the rattlesnake belt of Ontario, yet that does not prevent people from travelling by motor car, and enjoying it. Motor vehicles caused 1,180 deaths in Ontario in a single year (1956) while the massasauga killed one person here in the current century. Comparing these risks, one should find it easier to think sensibly about the massasauga. Although there appears to be no rule of safety that will afford you protection from the risk of motor death, there is one that will almost certainly protect you from massasauga bite, and that is, *to learn to recognize the snake and leave it alone.*

The young, usually seven or eight in number, are born from late July to early September.

DISTRIBUTION IN ONTARIO. This snake is found along the Lake Erie and Lake Huron shores, and at Georgian Bay, including the larger islands. Its present range extends inland from these waters for about twenty or thirty

Baby Massasauga, showing first (terminal) button of future rattle

miles, its numbers thinning out. In former years it had a wider range and was far more plentiful than it is now. It is still moderately common in some restricted areas, as on the Bruce Peninsula. It appears to be absent from Manitoulin Island, and from the northern shore of the mainland west of Killarney.

TIMBER RATTLESNAKE

(Banded Rattlesnake)

SIZE AND STRUCTURE. The timber rattlesnake is a large, thick-bodied snake, occasionally reaching a length of five feet. The head is broad and markedly triangular, and the neck narrow. There are many small, irregular scales on top of the head between the eyes. There is, of course, the facial pit between the eye and nostril, the vertical eye pupil, and a rattle terminating the tail.

COLOUR. The ground colour above varies from yellowish to gray or brown, sometimes black. It is marked with a series of dark, more or less V-shaped crossbands, with the arms of the Vs directed forward. The tail is black or banded. The under surface may be white, yellowish or buff, and more or less flecked with gray or brown.

HABITS AND HABITAT. The preferred home of the timber rattlesnake is in more or less wooded situations with rocky ledges, where deep, secure and dry hiding places may be found. Its former range in southern Ontario, when it was more common, apparently coincided with such habitat.

Although it is a large and powerful snake with long fangs, and a good supply of venom, it is reported by authorities in the eastern United States, where it is common, to be of rather quiet temper, preferring escape to conflict with man. Nevertheless, it is a dangerous snake and takes a small toll of human lives. In the United States, in the eight-year period from 1927 to 1934, it inflicted 215 known bites, and 14 of the

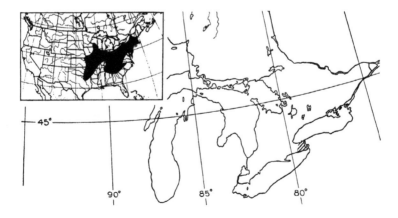

people bitten died. One death of a soldier near Queenston in 1812 was probably caused by this snake. We know of no other Canadian fatality from it, nor indeed, of any other case of its bite in this country.

This rattlesnake feeds almost entirely upon mammals, chiefly rodents.

DISTRIBUTION IN ONTARIO. Found only in the vicinity of Niagara gorge, it is now nearly extinct in the Province. It formerly had a wider range, following the limestone outcropping to as far north as Manitoulin District, where a specimen was caught on Fitzwilliam Island in 1886. There is an unmistakable reference to this snake having been seen by M. de la Salle, while he was climbing the rocky escarpment near to the little lake now known as Lake Medad, in Halton County. This was in the year 1669, and is the earliest positive reference to the timber rattlesnake in Ontario. It gives us a hint of its former range along the rocky ridge.

A specimen killed at Point Pelee in 1918 had probably drifted over from the western Lake Erie islands, on some of which a few still survive, or had survived until quite recent years.

BOOKS

THE BOOKS LISTED BELOW are clearly written, and all are illustrated. They were prepared in order to be useful to both the beginner and the advanced student, the layman and the professional. They all contain some technical information, but the text is not heavy.

There are many other good books dealing with snakes and other reptiles, references to which will be found in the bibliographies of the books listed below.

BARBOUR, THOMAS. *Reptiles and Amphibians.* Houghton Mifflin Co., 1926. Pp. 125.

CURRAN, C. H., and KAUFFELD, C. *Snakes and Their Ways.* Harper Bros., 1937. Pp. 285.

DITMARS, RAYMOND L. *Snakes of the World.* The Macmillan Co., 1931. Pp. 207.

———— *The Reptiles of North America.* Doubleday, Doran, 1936. Pp. xiv, 476.

LOGIER, E. B. S. *The Reptiles of Ontario.* Royal Ontario Museum of Zoology Handbook, no. 4, 1939. Pp. 63, ii.

OLIVER, JAMES A. *The Natural History of North American Amphibians and Reptiles.* D. Van Nostrand Company Inc., 1955. Pp. xi, 359.

POPE, CLIFFORD H. *Snakes Alive and How They Live.* The Viking Press, 1937. Pp. xii, 238.

———— *The Reptile World.* Alfred A. Knopf, 1955. Pp. 325, xiii.

SCHMIDT, KARL P., and DAVIS, D. D. *Field Book of Snakes of the United States and Canada.* G. P. Putnam's Sons, 1941. Pp. xiii, 354.

WRIGHT, ALBERT HAZEN, and WRIGHT, ANNA ALLEN. *Handbook of Snakes of the United States and Canada.* Comstock Publishing Associates, Cornell University Press, 1957. Pp. xvii, 1105, 2 vols.

APPENDIXES

SYNOPTIC LIST OF ONTARIO SNAKES

Family COLUBRIDAE

QUEEN SNAKE: *Natrix septemvittata* Say

NORTHERN WATER SNAKE: *Natrix sipedon sipedon* Linnaeus

ISLAND WATER SNAKE: *Natrix sipedon insularum* Conant and Clay

DEKAY'S SNAKE: *Storeria dekayi dekayi* Holbrook

RED-BELLIED SNAKE: *Storeria occipitomaculata occipitomaculata* Say

BUTLER'S GARTER SNAKE: *Thamnophis butleri* Cope

EASTERN RIBBON SNAKE: *Thamnophis sauritus sauritus* Linnaeus

EASTERN GARTER SNAKE: *Thamnophis sirtalis sirtalis* Linnaeus

RED-BARRED GARTER SNAKE: *Thamnophis sirtalis parietalis* Say

EASTERN HOG-NOSED SNAKE: *Heterodon platyrhinos platyrhinos* Latreille

EASTERN RING-NECKED SNAKE: *Diadophis punctatus edwardsi* Merrem

EASTERN BLUE RACER: *Coluber constrictor flaviventris* Say

EASTERN SMOOTH GREEN SNAKE: *Opheodrys vernalis vernalis* Harlan

EASTERN FOX SNAKE: *Elaphe vulpina gloydi* Conant

BLACK RAT SNAKE: *Elaphe obsoleta obsoleta* Say

EASTERN MILK SNAKE: *Lampropeltis doliata triangulum* Lacépède

Family CROTALIDAE

EASTERN MASSASAUGA: *Sistrurus catenatus catenatus* Rafinesque

TIMBER RATTLESNAKE: *Crotalus horridus horridus* Linnaeus

THE SCIENTIFIC NAME

PEOPLE OFTEN ASK, why use Latin names in zoology? Are not English names just as good, and easier? There are two reasons for using Latinized scientific names for animals (and plants): one is to express the natural relationship of the creature; the other is to establish a universal name for each, that will be exactly the same in every country and every language. Let us deal first with natural relationship.

Biologists believe that all life is related by descent, and that the biological relationships of living things are revealed in their structure, physiology and processes of development. On this basis the animal kingdom is distinguished from the vegetable kingdom, and in each of these kingdoms a few large groups, called phyla (singular, phylum) have been recognized, and sub-divided into successively smaller groups called subphyla, classes, orders, families, genera (singular, genus), species and races. The members of one phylum are related together by some very fundamental characters in which they all agree. The members of the different classes in a phylum all possess the fundamental characters of the phylum, but also possess other characters of their own which relate them together, but separate them from the members of any other class in the phylum, and so on down through orders, families, genera, and species, to races.

It was pointed out on the first page of this book that fish, amphibians, reptiles, birds and mammals, though differing so greatly in appearance, are all related to each other by their common possession of a backbone. They are therefore known as vertebrates and placed in the subphylum Vertebrata, of which they form five separate classes.

I will now set down the complete classification of one of the snakes in the synoptic list, the eastern milk snake.

> PHYLUM: Chordata—all animals in which a notochord appears during the course of development.
> SUBPHYLUM: Vertebrata—all animals possessing a backbone.
> CLASS: Reptilia—all reptiles.
> SUBCLASS: Synaptosauria—lizards and snakes.
> ORDER: Serpentes—all snakes.
> FAMILY: Colubridae—all snakes possessing certain definite structural features in their skulls, not found in other snakes.
> GENUS: *Lampropeltis*—the king snakes (including the milk snakes).
> SPECIES: *doliata*—the milk snakes.
> RACE: *triangulum*—the eastern milk snake.

The names of the genus, species and race are the only ones used in the scientific name, but that is sufficient to state the zoological relationship of the animal. In the above case, the species name *doliata* cannot be used for any other species within the genus *Lampropeltis*. The genus name, likewise, can be used only for the one single generic group of snakes belonging to the family Colubridae, and so on, back to the phylum. It follows then that when the genus and species names are written together, the zoological position of the animal is stated. Thus, *Lampropeltis doliata* written together signify a milk snake, and nothing else; if the third name *triangulum* is also written, it signifies the *eastern* milk snake as distinct from any other races of milk snakes.

Now, for the second reason, that of universal names. The Latinized names used above for all the groups from phylum to race are written, spelled, and pronounced in the same way in every language. Thus, the Latinized words *Lampropeltis doliata* are the same in any language, that is, they are universal words. That would not be true of the words "milk snake" which would have to be translated into different words in every language; so they would be different in English, French, German, Spanish, or whatever language you care to name.

Since the generic and specific names are always written together in the Latinized scientific name, it both expresses zoological relationship, and is understandable in every language of the civilized world.

It is customary to write generic, specific, and racial names in italics, and to follow this with the name of the authority for the specific or racial name. Thus for the eastern milk snake, the first zoologist to apply the name *triangulum* to a milk snake was Lacépède, and the kind of milk snake to which he applied it was the eastern milk snake, therefore, the full scientific name of the eastern milk snake is *Lampropeltis doliata triangulum* Lacépède. Generic names, and group names above genus, as family, order, etc., are always capitalized; species and race names are not capitalized.

It is a common practice in writing, after the scientific name has been stated, to refer to it thereafter in the same article by the use of the capitalized initial of the genus, followed by the specific name, as *L. doliata*, or if the racial name, too, has been stated, to use only the initial letter for the species also, as *L. d. triangulum*. This is done to conserve space.

In the early days of the history of science it was the custom to write scientific treatises, and descriptions of animals, in Latin, which being a "dead" language does not change as do "live" languages. Therefore, a statement written in Latin always means, everywhere and at every time, precisely the same thing that it meant when and where it was written.

Carolus Linnaeus, the Swedish botanist, who lived from 1707 to 1778, and originated the *binomial* system of naming living things, by genus and species, wrote his works in Latin.

66

Let us now take a closer look at the species, and race or subspecies, the groups involved with the genus in the scientific name. A species is a group of individuals (animals or plants) which possess some distinctive character, or characters, in common and freely interbreed, reproducing these characters in their young. Related species which differ from each other only in small degrees or minor features are grouped together in one genus, the next superior group. The different species of garter snakes, which are closely related to each other, are placed together in one genus, *Thamnophis*; they are more closely related to each other than they are to the king snakes, for instance, which are placed together in another genus, *Lampropeltis*. Separate species, even when related, do not *usually* interbreed, even when co-occupying the same territory.

When a species extends its range over a large area many hundreds of miles across, so that different sections of the species population are prevented by distance or other barriers from freely interbreeding, differences gradually appear between them. When the differences have developed to the point where they serve to distinguish and identify these different sectional populations of the species, they are then recognized as subspecies or races and given scientific names. The race name thus expresses a geographical relationship within the species, and it is often necessary to know precisely from where a specimen came before it can be identified to race.

Since the races of a species are capable of interbreeding when they meet, and do so, it follows that two or more races of one species can never occupy the same geographic area and maintain their identity; any differences between them would be dissolved out by interbreeding and the population would become homogeneous (of the same content throughout). In the belts where two or more such races come together interbreeding does occur, and the population there becomes—or remains—intermediate between the several races or subspecies, and is known as intergrade. It is usually impossible to assign individuals from such an area to one or another particular race.

APPENDIX 3

KEY TO ONTARIO SNAKES

HOW TO USE THE KEY. This key is based upon external characteristics that can be seen easily. It will work for the assemblage (geographical group) of snakes found in Ontario (and for all of eastern Canada if you mentally add a sub-clause under KK, reading "uniform black above: BLACK RACER *Coluber constrictor constrictor*"). It will not work for snakes found west of Ontario.

The key is divided into successively inferior sections, each indexed with a capital letter. To each of these sections there is an alternative of equal rank indexed with the same letter doubled. To use the key, proceed as follows. Compare your specimen with the first alternative, A, and if it does not agree with this, go on to the second alternative, which is AA. If it does agree with alternative A, compare it with the next statement under that section, which is B, and continue in this manner as long as the specimen agrees with each successive statement. If you come to a statement with which it does not agree, go on to the second alternative of this, which will be marked with the same letter doubled, and then continue as before so long as the specimen agrees with each successive statement, until you are led to a specific name.

The key is the standard instrument used in systematic books for the identification of such things as plants, animals, rocks and minerals. There are many ways of setting up a key; I used that which seemed to be the simplest for dealing with a small group.

With a little practice, one soon learns to use a key with ease and speed.

KEY TO ONTARIO SNAKES

A. No rattle on end of tail (Fig. 2I, J); no pit between eye and nostril (Fig. 2A, B, C).
B. Snout rounded, not pointed and protruding (Fig. 2A, C).
C. Anal plate divided (Fig. 2H).
D. Keels present on some or all of dorsal scales (Fig. 2E).
E. All scales, at least above first row, strongly keeled (Fig. 2E).
F. Scales in fewer than 19 rows; no loreal plate (Fig. 2A).
G. Scales in 15 rows; a yellowish spot on back of neck; belly usually red: RED-BELLIED SNAKE, *Storeria occipitomaculata occipitomaculata* (Plate II A, pages 30–1).
GG. Scales in 17 rows; no yellowish spot on back of neck; belly brownish white or pinkish: DEKAY'S SNAKE, *Storeria dekayi dekayi* (pages 28–9).
FF. Scales in 19 or more rows; loreal plate present (Figs. 1A, 2C).
H. Scales in 19 rows; uniform brown above, or with 3 narrow black stripes; a yellow lateral stripe on adjacent halves of first and second scale rows: QUEEN SNAKE, *Natrix septemvittata* (pages 24–5).

68

HH. Scales in 23 or 25 rows.
I. Brown above, usually blotched; no lateral light stripe; ventral plates usually dark mottled, often with dark-edged half-circles: WATER SNAKE, *Natrix sipedon sipedon* (pages 26–7).
II. Gray above, blotches absent or indistinct; ventral scutes white, unmottled: ISLAND WATER SNAKE, *Natrix sipedon insularum.*
EE. Scales of first three to five rows smooth or nearly so, others weakly keeled.
J. Brownish yellow above with large dark blotches: FOX SNAKE, *Elaphe vulpina gloydi* (pages 46–7).
JJ. Uniform black above, or obscurely blotched: BLACK RAT SNAKE, *Elaphe obsoleta obsoleta* (pages 48–9).
DD. All dorsal scales smooth, no keels (Fig. 2G).
K. Scales in 15 rows, length less than 24 inches.
L. Slate gray above; belly yellow or orange, a ring of the same colour around neck: RING-NECKED SNAKE, *Diadophis punctatus edwardsi* (pages 40–1; Plate II B).
LL. Colour above grass green; no ring around neck: SMOOTH GREEN SNAKE, *Opheodrys vernalis vernalis* (pages 44–5; Plate II C).
KK. Scales in 17 rows; uniform bluish green above; length up to 6 feet; young blotched: BLUE RACER, *Coluber constrictor flaviventris* (pages 42–3).
CC. Anal plate not divided (Figs. 1D, 2F).
M. All dorsal scales smooth, no keels, in 21 or more rows; large black-edged, brown dorsal blotches: MILK SNAKE, *Lampropeltis doliata triangulum* (pages 50–1).
MM. All scales, at least above first row, strongly keeled, in 19 rows; normally with three longitudinal light stripes.
N. Lateral stripe anteriorly on third, and involving adjacent halves of second and fourth, scale rows, a light spot on preocular: BUTLER'S GARTER SNAKE, *Thamnophis butleri* (pages 32–3).
NN. Lateral light stripe anteriorly on third and fourth scale rows; tail 0.28 or more of total length; a light spot on preocular: RIBBON SNAKE, *Thamnophis sauritus sauritus* (pages 34–5).
NNN. Lateral stripe on second and third scale rows, sometimes ill-defined below, absent in black specimens; tail 0.25 or less of total length.
O. If dorsolateral spots visible, those of upper row fused together, interspaces of lower row not red: EASTERN GARTER SNAKE, *Thamnophis sirtalis sirtalis* (pages 36–7).
OO. If dorsolateral spots visible, those of upper row not fused together, interspaces of lower row red: RED-BARRED GARTER SNAKE, *Thamnophis sirtalis parietalis.*
BB. Snout pointed and protruding (Fig. 2B), keeled above; head broad, body thick; usually blotched above, sometimes uniform olive, or black: HOG-NOSED SNAKE, *Heterodon platyrhinos platyrhinos* (pages 38–9).
AA. Tail terminating in a rattle (Fig. 3E); a deep pit between eye and nostril (Fig. 3B); eye pupil a vertical slit (Fig. 3B).
P. Top of head with large symmetrical plates (Fig. 3A); dorsal pattern of blotches: MASSASAUGA, *Sistrurus catenatus catenatus* (pages 56–9).
PP. Top of head with small, unsymmetrical scales (Fig. 3B, C, D); dorsal pattern of crossbands, or black: TIMBER RATTLESNAKE, *Crotalus horridus horridus* (pages 60–1).

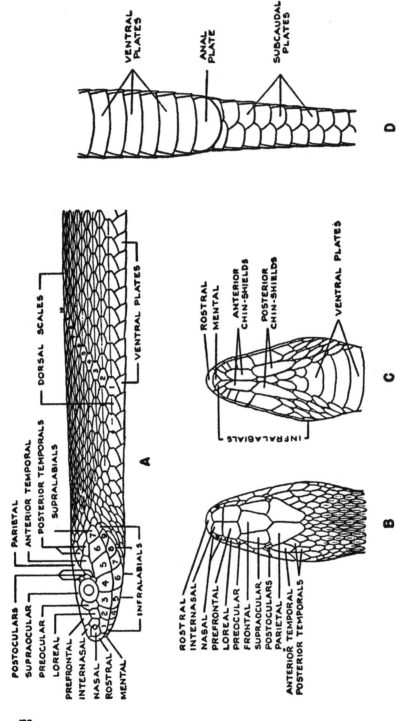

FIGURE 1: A, side view of head and forward part of body of snake showing scales; B, top view of head; C, under-surface of head; D, under-surface of anal region showing ventral, anal and subcaudal plates

70

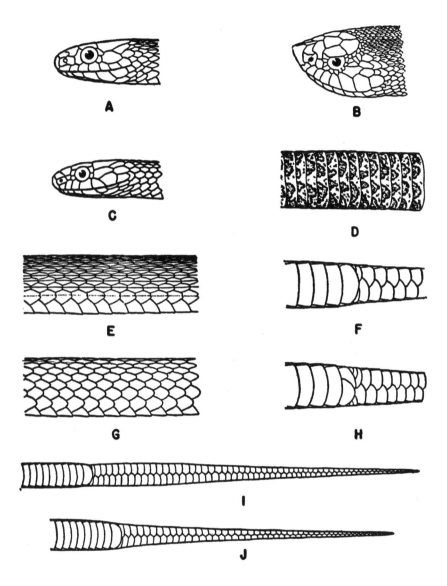

FIGURE 2: A, head of DeKay's snake showing two nasal scales with nostril
between, no loreal plate; B, head of hog-nosed snake showing round eye
pupil and pointed, protruding snout; C, head of smooth green snake
showing loreal scale; D, ventral view, middle of body of water snake,
showing blotched colour pattern; E, keeled scales; F, undivided anal plate;
G, smooth (unkeeled) scales; H, divided anal plate; I, tail of male snake,
swollen at base; J, tail of female snake, tapering at base

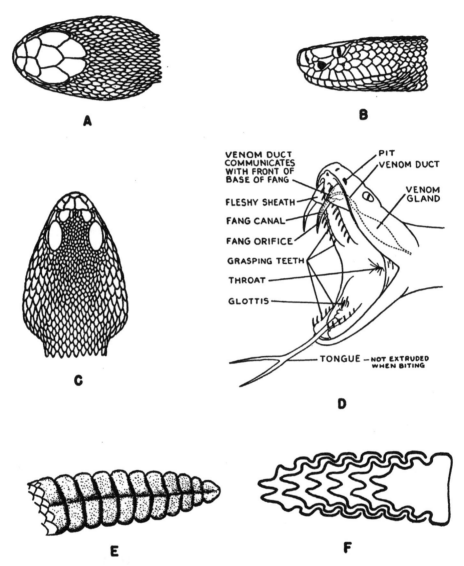

A

B

C

VENOM DUCT
COMMUNICATES
WITH FRONT OF
BASE OF FANG

FLESHY SHEATH

FANG CANAL

FANG ORIFICE

GRASPING TEETH

THROAT

GLOTTIS

PIT
VENOM DUCT

VENOM
GLAND

TONGUE – NOT EXTRUDED
WHEN BITING

D

E

F

FIGURE 3: A, head of massasauga, top view, showing large symmetrical scales between eyes; B, head of massasauga, side view, showing pit and vertical eye pupil; C, head of timber rattlesnake, top view, showing small irregular scales between eyes; D, venom apparatus of rattlesnake; E, rattle of timber rattlesnake; F, section through basal portion of rattle, showing interlocking attachment of rattle segments

DETERMINATION OF SEX

IT IS OFTEN desirable to know whether a particular animal is male or female. For adult snakes, this point can often be decided from external features. In females the basal part of the tail is narrower and more tapering than in males (Fig. 2J). In males it is fuller, only slightly or not at all tapering, or even somewhat swollen (Fig. 2I). This fullness is due to the *hemipenes*, or paired copulatory organs, which lie side by side in this part of the tail. They may be exposed to view in dead specimens by making a short incision lengthwise behind the vent; in females only the scent glands and muscle tissue will be seen. In young specimens it may be impossible to determine the sex in life, but in dead specimens it may usually be done by the simple dissection described above.

KEEPING SNAKES AS PETS

MOST OF OUR SNAKES may be kept as pets. A cage of suitable size and design is necessary, but it need not be elaborate or expensive. For small snakes a cage with 1 × 2 feet of floor space will serve; 2 × 4 feet, or preferably larger, for our largest snakes; the height is less important, a foot for the small cages and from two to three feet for the large ones is suggested. They may be built of any seasoned wood, waste lumber or plywood, with a glass front and a hinged screen wire top; or they may be all of glass (except the screen top) with a wood or metal frame to support the glass. I have often made cages out of wooden grocery boxes or butter boxes in the days when they were easily obtainable. Before using the cage, give the wooden parts several (two or three) coats of a good grade of waterproof paint or varnish, and allow to dry and air until all trace of paint odour has gone.

For bedding, paper, dry grass, leaves or moss are satisfactory; paper below and dry leaves or moss on top is a good arrangement. Snakes must have cover. A water dish or jar large enough for the snakes to get into and soak themselves should be provided. The water should be kept fresh and clean (renewed every one or two days), and the bedding renewed when it becomes soiled, too damp, or stale. Cleanliness is important if the snakes are to be healthy.

Most snakes enjoy direct sunlight if not too hot, but there should always be a shaded half of the cage where they can get out of the sun, which will quickly kill them if too hot.

The food required by the different species will be found under the accounts of the species. How often to feed will depend upon the kind of snake and quantity of food. Garter snakes, water snakes, DeKay's snakes and red-bellied snakes will usually feed two or three times a week, or every day if the meals are small. The constricting snakes will do on one meal a week, if sufficiently large. One large mouse for an adult milk snake, and two or three for a fox snake. In general, snakes should be fed as often as they will eat, but they will come to no harm if they have to fast for a few days at any time.

Captive snakes sometimes have difficulty in shedding their skin. A skin that should be shed has a dry, dull, and sometimes wrinkled appearance. About ten days or so before shedding, a snake's eyes become whitish and milky looking; this condition lasts for three or four days, and then the eyes clear. About a week after the eyes clear the skin should be shed, and if this does not happen naturally, the snake will require some help. Soak it in tepid water (about 72° to 80°F.) for two or three hours; loosen the skin at the tip of the snout and chin and at the edges of the lips, and roll it backwards, inside out, being careful that the scutes over the eyes come with it. The old skin must be worked back slowly, and the snake will usually assist by muscular movements of its skin. After shedding a snake is usually hungry, for it often does not feed from the time when the eyes cloud over.

Climbing snakes, like the fox snake, black rat snake and milk snake, like to have a piece of tree limb in the cage, upon which they may climb and rest. With a little practice, you will very soon learn how to take care of your snake pets properly.

COLLECTING AND PRESERVING OF SPECIMENS

REFERENCE COLLECTIONS of properly preserved specimens are of great importance to zoologists. The systematic study of snakes and their distribution depends upon such collections, which are maintained by museums and other institutions that carry on such work. Good preservation is necessary; poorly preserved specimens may be very difficult to identify or study because important features may be distorted, obscured, or lost. If you are interested in the study of snakes, you may wish to build up a reference collection of your own, or to contribute specimens to a museum that is doing such work.

If you wish to collect specimens, you will find information on where to look for them in the accounts of the species in the preceding pages.

Rattlesnakes should not be collected by inexperienced people. They must be picked up with a snake hook or slip noose at the end of a stick of sufficient length (three or four feet). There is always risk in handling them, and even well-experienced people have been severely bitten. Leather gloves may be desirable when collecting the larger harmless snakes because some species will bite savagely when first caught.

Cotton bags are the most satisfactory field containers for carrying specimens. Sugar or flour bags (rehemmed in a sewing machine if the edges are frayed) are very good, or inexpensive bags may be made of factory cotton. A stout tie-string about twenty-four inches long should be attached by its middle about two or three inches below the top of the bag. Bags containing snakes must be tied tightly.

The specimens should not be crowded in the bags, or very small snakes carried in the same bag with very large ones, or injury may result. Bags containing snakes *should not be left lying in the sun.*

KILLING. Snakes may be killed with anaesthetics or by drowning. For the latter method enclose the specimens in a jar, perforated can, or weighted cloth bag and submerge in cool or tepid water, taking care to liberate all air bubbles. Drowning may require several hours. The quickest, most humane and convenient method of killing is by anaesthesia with the vapour of chloroform, ether, or carbon tetrachloride. Carbon tetrachloride* is the most satisfactory because it leaves the specimens limp in death. To use, place the specimens in an air-tight can or jar with a wisp of cloth or paper moistened with the fluid. Snakes will be unconscious in from five to fifteen minutes, and will die in

*Dangerous poison: do not spill on hands or inhale vapour.

about thirty or forty-five minutes. The vapour of dichloricide (a common moth killer) will quickly kill snakes if they are enclosed in a jar with the crystals.

The common practice of killing specimens in the preserving solution of formalin or alcohol is a harsh and cruel method, for which there is no excuse. No creature should be tortured to death, and other means can always be found.

PRESERVATION. To preserve a specimen, inject the body cavity with 10 per cent formalin (one part commercial formalin to nine parts of water), and submerge in a solution of the same strength, straightening or coiling the body into the desired position, for it cannot be changed afterwards. Injection should be done at several points along the belly, and into the base of the tail. If a hypodermic syringe is not available, make a series of incisions (with a small scissors), about an inch long and an inch apart, along the belly, and one or two in the basal part of the tail, to admit the preserving fluid. Knead the specimen a little to remove air bubbles and make sure the fluid enters.

Specimens may be left in formalin indefinitely, but it is better to transfer them to alcohol after a few days, for permanent storage. The alcohol should be of 75 or 80 per cent strength (two parts of water to eight parts of alcohol is close enough). If pure alcohol cannot be obtained, a good grade of denatured alcohol, methyl, or wood alcohol will serve. The latter alcohols may cloud upon being mixed with water, but will usually clear later by precipitation.

LABELLING. A label bearing the locality, date, and collector's name should be tied to each specimen before it is put into the preservative. The labels should be of strong linen paper and written with graphite pencil or india ink (if india ink is used, it must be *thoroughly dry* before immersion). Never use ordinary writing inks, for they soon wash out.

CONTAINERS. Glass jars are best for storing small specimens; the tops should seal tightly. Large specimens can be stored in earthenware crocks, which may be sealed fairly well against evaporation by laying several sheets of plastic, such as vinylite, over the tops of the crocks, underneath the lid. Wooden pails with tightly fitting lids are good field containers.

SHIPPING. When packing specimens for shipment, it is well to wrap each one in a piece of rag or cheesecloth to protect it and its label, and pack enough soft material such as rags, moss, excelsior, or paper between them to prevent rubbing, and to hold moisture. The packing material should be saturated with the preserving fluid, and the surplus then drained off. Plastic bags, tied tightly to prevent seepage, are good shipping containers and may be enclosed in paper cartons for transportation by express or mail.

GLOSSARY

MANY OF THE WORDS defined below have already been explained where they occurred in the text, but it seemed advisable to list them alphabetically with their meanings for quick reference.

AGGRESSOR. One who attacks first, unprovoked.

ANAESTHETIC. A substance, commonly a gas or fluid, which deadens feeling or causes insensibility.

ANAL. Pertaining to the anus or posterior opening of the bowel; situated at or in the anus.

ANAL PLATE. The scale immediately in front of the anus in snakes (Fig. 1D).

ANTERIOR. Toward the front, forward in position.

ANTIBIOTICS. Substances produced by certain bacteria or fungi that prevent the growth of disease-producing germs.

ANTISEPTIC. A chemical agent which checks the growth of disease-producing germs. Methods of procedure (as in surgery) which are opposed to the development of infection.

BACTERIA. Microscopic plants, various species of which are capable of causing disease or decay.

BACTERIAL. Pertaining to bacteria or their action.

BELLY. In a reptile, the under surface of the body between the head and the tail.

BIOLOGICAL. Pertaining to life or living processes, or the products produced by living things.

CALORIE. A unit of heat; the amount of heat that will raise the temperature of one gram of water one degree Centigrade (under a pressure of one atmosphere).

CARNIVOROUS. Flesh-eating. A term applied to animals that kill and eat other animals.

CARNIVORE. A flesh-eating animal.

CARTILAGE. A tough, elastic, more or less transparent material developed as a supporting tissue in the bodies of vertebrate animals, especially in the embryo and young.

CASUALTY. Injury or death (usually death) from an accident; the person so injured or killed.

CAUTERIZE. To sear with heat or some strong chemical.

CELL. The unit of living matter; the simplest unit of organization that can carry on living functions and reproduce itself. The minute structures of which the bodies of animals and plants are composed.

CLASS. One of the main subdivisions of a phylum.

COLD-BLOODED. A term applied to animals whose temperature varies with that of their environment; ectothermic.

COMA. A state of insensibility brought on by a poison or injury.

CONCAVITY. A hollow or depression.

DISINFECTANT. A substance that kills bacteria or other disease-producing germs.

DORSAL. Pertaining to the back; situated on or in the back.

DORSAL SCALES. The scales on the back and sides of a snake's body above the ventrals (Figs. 1A, 1D).

DORSOLATERAL. Pertaining to the side or sides of the back, or situated thereon.

ECTOTHERMIC. Dependent upon external temperature; obtaining the heat required for bodily processes from outside the body.

EMBRYO. An animal in the early stages of development prior to hatching or birth.

EMBRYONIC. Pertaining to an embryo; a term applied to the developmental stages undergone within the egg or the body of the mother.

ENDOTHERMIC. Capable of maintaining a constant body temperature independently of that of the environment.

ENVIRONMENT. The kind of surroundings in which a creature lives, or happens to be.

EVOLUTION. The process of development of races, species, or larger groups of living things by a series of changes from earlier and more primitive kinds— as reptiles from amphibians. The biological theory that living species had their origin in pre-existing and different forms of life.

EXOTIC. Foreign, belonging to a foreign country. Not native.

FAMILY. The principal subdivision of an order; the category above genus.

FANG. A long, sharp, spike-shaped tooth (Fig. 3D).

FORMALIN. A preserving fluid and disinfectant composed of formaldehyde gas dissolved in water.

GENUS. A group of related species within a family.

GLOTTIS. The opening, in the floor of the throat or mouth, into the wind pipe (Fig. 3D).

HABITAT. The kind of place in which a creature lives; its natural abode, as in water, on land, in a forest, on a prairie, etc.

HAEMORRHAGE. A profuse leakage of blood from broken blood vessels.

HERPETOLOGIST. One who studies reptiles and amphibians.

HIBERNATION. The winter sleep of some kinds of animals, in which the body temperature is lowered and its physical and chemical processes are slowed down.

HOMOGENEOUS. Of the same content and consistency throughout.

INCISION. A cut made with a sharp edge.

INSTINCT. Hereditary behaviour; a tendency to behave in a certain way that leads to the attainment of some natural goal of the species.

INVERTEBRATE. An animal without a backbone.

JACOBSON'S ORGAN. A pair of sensory cavities in the roof of a snake's mouth, near the front, which function as organs of smell.

KEELED. Possessing a ridge or keel; said of scales that have a ridge down the centre (Fig. 2E).

LATERAL. Pertaining to the side; situated on the side or sides.

LIGAMENT. A band of tough connecting or supporting tissue in an animal's body.

LOREAL SCALE. A scale situated between the nasal and preocular scales (Fig. 2C).

MAXILLARY BONE. The outer, lateral, tooth-bearing bone of either side of the upper jaw, shortened in vipers and bearing only the fangs (Fig. 3D).

MICRO-ORGANISM. A living thing of microscopic size.

MIDDLE EAR. That portion of the ear between the eardrum and the inner ear, containing a chain of movable bones that convey sound vibrations inward from the eardrum.

NAUSEA. A feeling of stomach sickness, a tendency to vomit.

NOCTURNAL. Pertaining to the night; active at night.

NOTOCHORD. A dorsal rod of cartilage which appears very early in the embryonic development of all members of the phylum Chordata. In vertebrates, the vertebral column or spine develops around it.

ORDER. A subdivision of a class, above a family.

ORGANISM. A term applied to living things—whose structure is organized so as to carry on the functions of life.

PALLOR. Paleness, loss of natural colour from the face.

PERPENDICULAR (TO). At right angles to.

PHYLUM. One of the main divisions of the animal (or vegetable) kingdom.

PHYSIOLOGY. The life processes of a living body, such as circulation, etc.; or the study of them.

PIT. A temperature-sensitive cavity in the side of the face, in front of the eye, in rattlesnakes and certain related vipers, which are therefore known as Pit Vipers (Figs. 3B, 3D).

POSTERIOR. Behind, toward the back, backward in position.

PRECIPITATION (IN A FLUID). The separating out and sinking down of some contained substance.

PREHENSILE. Adapted for grasping.

PREOCULAR. Situated in front of the eye; the scale (or scales—preoculars) immediately in front of the eye in snakes (Fig. 1A).

PRIMITIVE. Pertaining to the beginning or origin; said of animals that still retain ancient characters belonging to an early stage of their evolution.

PROTOZOA (*Sing.* protozoan). Simple, usually single-celled and microscopic animals, belonging to the phylum Protozoa.

RACE. A subdivision of a species; same as subspecies.

RODENT. A gnawing animal, as a mouse, rat, squirrel or rabbit, etc.; any member of the order Rodentia.

RECEPTORS. Special organs, cells, or nerve endings that receive the stimuli that produce sensations such as those of heat, cold, pain, smell, etc.

SPECIALIZED. Developed and restricted to serve a particular purpose.

SPECIES. A group of related individuals within a genus, that possess some distinctive character, or characters, in common and do or may freely interbreed, reproducing those characters in their young.

SUBCAUDALS. The scales on the under surface of the tail (Fig. 1D).

SUBPHYLUM. The most important subdivision of a phylum.

SUBSPECIES. A subdivision of a species; the same as race.

SURGICALLY STERILE. Completely devoid of bacteria or anything that could cause infection. Many things that appear to the eye to be perfectly clean are not surgically sterile until treated with heat or a disinfectant.

TAIL. That portion of the body behind the vent.

TRANSVERSE. Lying crosswise on the body; at right angles to the long axis.

VENT. The anus or posterior opening of the bowel.

VENTRAL. Pertaining to the belly; situated on the belly or on the lower surface of an animal's body.

VENTRAL SCALES OR PLATES. The (usually) broadened scales on the lower surface of a snake's body between the head and the vent.

VERTEBRATE. An animal with a backbone or spine.

WARM-BLOODED. A term applied to animals that maintain a constant body temperature independently of that of their environment; endothermic.

RATTLESNAKE BITE

IT WOULD probably be a gross overstatement to say that even one person in any
five thousand picked at random would know what to do in a case of rattlesnake bite. Erroneous and harmful ideas on the subject are widely held, and
some of them are even advocated in first aid textbooks of fairly recent years,
and in magazine articles, upon which many people rely. In Canada, rattlesnake
bite is very rare; so the general ignorance about proper treatment is naturally
great. This ignorance exposes a bitten person to a greater than necessary risk.

Therefore, it seemed proper to close this little book on the snakes of
Ontario with an account of the proper first aid treatment for snake bite, in
accord with modern knowledge and methods, for the guidance of campers,
teachers, and senior pupils. I have also included some notes addressed to the
physician or hospital staff who may be called upon to treat a case of rattlesnake bite. What is said below applies to the bites of all kinds of rattlesnakes
found in Canada.

Prevention of Bite

WHEN IN RATTLESNAKE COUNTRY watch where you place your feet and, if the
terrain is reedy, bushy or rocky so that a rattlesnake might lie concealed,
wear protective clothing. Leather shoes, socks, and loose-fitting trousers,
slacks or jeans afford fair protection against the massasauga. When in territory
inhabited by larger rattlesnakes, the use of high boots or leggings is advisable. Canvas shoes and shorts are an invitation to trouble in rattlesnake
country. Learn to listen for the sound of the rattle, and if you hear it, stop,
and locate the snake before you take your next step. If you step back or to
one side without looking, you may step on it, or too close to it. Do not walk
outdoors at night without a flashlight. Do not poke your hands into places
where you cannot see. Be careful when gathering kindling, you may acci-
dentally grasp or pick up a snake. Do not gather kindling at night. Before
picking berries, beat about the bushes with a stick. If a rattlesnake is there,
this will disturb it and it will rattle.

In general then, use ordinary common sense. Watch where you put your
hands and feet; listen for the sound of the rattle; if you see a rattlesnake,
give it space and it will not harm you; use a flashlight at night; wear protective clothing when necessary. The fear of rattlesnakes need not spoil your
summer holiday anywhere in Ontario.

Symptoms of Rattlesnake Bite

I SHOULD STATE immediately that at the time of the accident it is impossible to assess the possible seriousness of any rattlesnake bite. A rattlesnake as large as an adult massasauga is fully capable of delivering a fatal bite, but for one or more reasons usually fails to do so. It is doubtful that the over-all death rate from untreated rattlesnake bites would much exceed 15 per cent. For massasauga bites, specifically, the death rate in untreated persons would be much lower. However, this statement is not intended to promote carelessness, but rather to allay unnecessary and harmful panic, and every bite should be treated as serious until proven not to be.

If a person is bitten by a rattlesnake there will be one or two fang punctures. Severe local pain usually develops at once, or within a few minutes, and swelling begins within about ten minutes. Purplish discolouration around the bite soon appears, and the swelling progresses up the bitten limb. Shock commonly follows soon with faintness, dizziness, nausea, pallor, perspiration and weak, quick pulse. The patient may vomit, and may have some difficulty in breathing, and pain in the region of the heart. In severe cases respiratory difficulty and heart pain may be acute. Internal haemorrhages may occur, with bleeding from the nose, mouth, kidneys, bladder and bowels, and the vomiting of blood-stained material. Loss of vision, paralysis, coma and death may follow.

A bitten person may faint soon after the bite, but this in itself is not necessarily serious; it may result from fright, even after the bite of a harmless snake. Nevertheless, fright *can* be serious when it involves much shock, and fainting is a symptom of shock. If consciousness does not return quickly, treat the condition by standard methods, but avoid the use of alcoholic stimulants.

The Bite of an Unidentified Snake

IF ONE should be bitten by a snake which then escapes, and there is doubt as to whether it was a rattlesnake, apply a tourniquet, as directed below, and then wait to see if local symptoms develop *before proceeding with further treatment*. If the snake was a rattlesnake, the local symptoms will soon appear, and you cannot mistake them. In the meantime, the tourniquet will afford sufficient protection against the spread of the venom, if any.

What *Not* to Do If Bitten by a Rattlesnake

DO NOT RUN or exercise unnecessarily. Do not use any alcoholic drink. Do not rub or inject permanganate of potash, carbolic acid, or any other chemical into the bite; do not cauterize with heat by any means whatever, or with any caustic. There is no "pocket" in which the venom can be reached and destroyed, and such treatments only damage the tissues further *and interfere with the subsequent removal of the venom.* Do not try any folklore "remedies" such as the flesh or entrails of a freshly killed chicken or other animal, a mud pack, tobacco poultice, kerosene, turpentine, camphor, ammonia, urine, etc. There are scores of such alleged "remedies," all totally useless, and a waste of valuable time. Some of them are positively harmful.

First Aid Objectives

THERE ARE four objectives to be obtained in the first aid treatment of rattlesnake bite: (1) to prevent the spread of the venom through the body, (2) to remove as much of it as possible from the region of the bite, (3) to neutralize as much as possible of that which is left in the body, (4) to prevent infection.

The standard and proven methods to accomplish these ends are: (1) the tourniquet, (2) incision and suction, (3) injections of antivenin. Infection must be avoided from the beginning in the first aid treatment by using surgically sterile procedure and antiseptic dressings; and in professional treatment by using antibiotics and injections against tetanus and gas gangrene.

First Aid Procedure

SEND FOR A DOCTOR, OR BRING THE PATIENT TO ONE, OR A HOSPITAL, BY CONVEYANCE. In the meantime, proceed with the following treatment.

TOURNIQUET. First and immediately place a tourniquet about two inches above the bite (between it and the heart). For a tourniquet, elastic rubber tubing or wide elastic tape are best, but a necktie, handkerchief, shoe lace, or strip of cloth torn from the clothing, or anything else, will serve. A wide tourniquet is better than a narrow one. Tie it in a loop and slip into position on the bitten limb, then slide a stick or pencil under it and tighten by twisting until a finger can just be forced under it, and secure in place. The purpose of the tourniquet is *to impede the flow of lymph* in which the venom is carried at first, but *not to stop the flow of arterial blood to the bitten part.* To do so is exceedingly dangerous. It should be possible to feel the pulse below the tourniquet. Loosen the tourniquet for about twenty seconds, not less, at fifteen-minute intervals (time it with a watch, don't guess). If the limb becomes numb or cold, the tourniquet is too tight; loosen and readjust. As the swelling advances up the limb, move the tourniquet up ahead of it, but close to it.

If medical aid can *certainly* be obtained within about two hours (in the case of massasauga bite), use only the tourniquet and leave subsequent treatment, including incision, to the doctor. However, *suction should be applied to the fang punctures at once.* Incision is better done under a local anaesthetic, and by a professional hand. If medical aid will not be available for more than two hours, proceed with incision.

INCISION. Sterilize the surface of the bite with some standard antiseptic, such as iodine or alcohol; sterilize a small, sharp knife or razor blade with one of these, or by holding it above, but close to, the flame of a match or lighter (avoid the deposition of soot). With it make a longitudinal cut at each fang puncture, one-eighth to one-quarter inch deep, and one-quarter to one-half inch long, but not more. If the bite is over a bone where the flesh is shallow, do not cut until the swelling develops, and then cut into the swelling only, but not to the bone. To lessen the risk of cutting blood vessels, nerves or tendons, *longitudinal cuts only* are recommended for the layman. If the snake was a very small one, omit incision.

If a blood vessel should be cut, arrest the bleeding by a small pressure pad (of moderate pressure) on the cut, and leave it until the bleeding has ceased. If bleeding persists after about twenty minutes, replace the pad, and make a substitute incision nearby. Unnecessary bleeding should be avoided. The purpose of the incision *is to extract lymph, not blood*; a snakebite victim needs all his blood.

SUCTION. Apply suction to the incised fang punctures, preferably by suction cup, but if necessary, by mouth. The venom is not poisonous in the mouth or stomach, so long as there are no breaks in the skin, or bad teeth, through which it could enter the lymph spaces or blood stream. If any pain develops in the mouth, stop mouth suction and use other means.* Wash out the mouth, preferably with some oxidizing agent such as a mild solution of permanganate of potash, or javel water (sodium hyperchlorite). The rubber cups or bulbs supplied in many snakebite kits are best, because they maintain gentle suction without the effort of an operator. They work better if applied to a wet surface. *Gentle suction is better than strong,* which tends to turn out the flesh at the incisions.

Additional small incisions, one-eighth of an inch long and one-eighth of an inch deep (or completely through the skin), should be made in the swelling around the bite, and also in that which has progressed up the limb—completely encircling it if necessary—and suction applied to as many of these as possible.

---

*Suction cups can be improvised with small bottles in the following manner. Light a match, thrust it into the bottle, and immediately apply the mouth of the bottle to the wet skin. The flame exhausts some of the air from the bottle, creating a partial vacuum, but is extinguished before it has time to appreciably heat the bottle; so there is no risk of a burn.

Suction should be continued through half of the first hour, and for at least twenty minutes of each subsequent hour, for up to fifteen or even thirty hours, for so long as there is poisoned lymph to drain away. When suction is not in process, cover the incisions with cloths wrung out of strong (preferably hot) epsom salt or common salt solution. This keeps them open and draining and helps to prevent infection from the outside.

ANTIVENIN. If antivenin is available, and it is apparent that a doctor or hospital cannot be reached for more than three (or up to four) hours, use immediately *after* applying the tourniquet, carefully following the directions which accompany the package (which should be memorized before an accident happens). *Before using, be sure to make the necessary tests for sensitivity to horse serum,* according to the directions, unless the patient is known not to be sensitive. If the patient is allergic to horse serum, the antivenin may cause death within a few minutes, or cause serious serum sickness. If the bite is on the face or trunk where a tourniquet cannot be applied, antivenin will have to be used as quickly as possible, but after desensitization if the patient is allergic (directions for desensitization accompany the package). This usually requires about half an hour.

All things considered, it is much better if possible, to have antivenin administered by a doctor, and in the case of massasauga bite, unless on the face, neck or trunk, or directly into a vein, there is no great danger in waiting for three or four hours for medical aid, provided that a tourniquet and suction have been used early. In the bites of larger rattlesnakes, as the Pacific, Prairie or Timber rattlers, prompter use of antivenin (within one or two hours) might be advisable.

Antivenin comes in 10 cc. packages, and for the bites of small rattlesnakes, one or two are usually sufficient. For the bites of larger snakes (as the timber rattlesnake) it is advisable to start the treatment with six or seven packages (60 or 70 cc.). The dosage will depend upon the size of the snake, the severity of the bite, and the weight of the patient. Children require larger doses than adults, up to twice as much, because their smaller bodies have less capacity to neutralize venom.

The first injection is usually given high up in the muscles of the bitten limb. It may also be injected into the swelling around the bite, and is very helpful there, but if there is only a limited supply available, withhold this local injection until suction has been used for about one and a half hours. Then inject and leave for an hour before resuming suction at that site. Antivenin treatment may have to be continued for many hours in severe cases.

Note carefully that antivenin *is not a substitute for the tourniquet, incision and suction.* Note also that it should not be used after the bite of a harmless snake, so in doubtful cases, follow the directions on page 81 under "the bite of an unidentified snake" before using it.

Antivenin may be obtained from John Wyeth and Brother, Walkerville, Ontario, or through their Toronto office, or through any druggist. However, the Ontario Department of Health has opened antivenin depots in twenty-five locations in the massasauga belt in Ontario; so no person bitten is likely, under ordinary circumstances, to be more than one or two hours distant from such treatment. The locations of these depots are as follows: *Barrie*, Royal Victoria Hospital; *Bracebridge*, Bracebridge Memorial Hospital; *Chatham*, Chatham Public General Hospital; *Collingwood*, Collingwood General and Marine Hospital; *Espanola*, Espanola General Hospital; *Goderich*, Alexandra and Marine General Hospital; *Hamilton*, Hamilton General Hospital; *Little Current*, St. Joseph's General Hospital; *London*, Victoria General Hospital; *Meaford*, Meaford General Hospital; *Midland*, St. Andrew's Hospital; *Orillia*, Orillia Soldiers' Memorial Hospital; *Owen Sound*, Owen Sound General and Marine Hospital; *Penetanguishene*, Penetanguishene General Hospital; *Parry Sound*, Parry Sound General Hospital, St. Joseph's General Hospital; *Port Colborne*, Port Colborne General Hospital; *St. Thomas*, St. Thomas-Elgin General Hospital; *Sarnia*, Sarnia General Hospital; *Southampton*, Saugeen Memorial Hospital; *Sudbury*, Sudbury District Health Unit; *Toronto*, Provincial Laboratory (Christie Street), Hospital for Sick Children; *Welland*, Welland County General Hospital; *Wiarton*, Bruce Peninsula and District Hospital; Windsor, City Health Department.

Further Care

IT IS ASSUMED that arrangements have already been made to obtain professional medical services as quickly as possible, or to move the patient to one of the above antivenin depots without delay. In the meantime keep the patient warm and still (the word "still" does not apply to the inevitable motion of a conveyance), for shivering or other muscular movement helps to distribute the venom through the body. Give him all the water he wants to drink, and hot tea or strong coffee are recommended. For faintness give a teaspoonful of aromatic spirit of ammonia in a glass of water. If possible, keep the bitten limb lower than the rest of the body, and keep it still, to promote drainage from fang punctures or incisions, and prevent the spread of the venom. Give aspirin or other mild drugs that may be on hand to relieve pain.

Keep the patient reassured; excitement and fear do positive harm, *fear alone has been known to cause death*, even after the bite of a harmless snake. With proper first aid treatment only, the chances of recovery are nearly 100 per cent, and if followed by professional medical care and antivenin, may be considered a practical certainty.

References

THERE IS a considerable literature on snakebite; however, the most easily available, complete, concise and plainly written single work on the modern treatment of snakebite in North America, that could be easily understood and followed by the layman, is: *Venomous Snakes of the United States and Treatment of Their Bites*, by William H. Stickel, Wildlife Leaflet 339 (United States Fish and Wildlife Service, Washington 25, D.C., 1952) 29 pages, illustrated.

An excellent account of modern methods of treatment, and a full medical bibliography, will be found in *Rattlesnakes, Their Habits, Life Histories, and Influence on Mankind*, by Laurence M. Klauber. University of California Press, 1956. Pp. xxix, 1476, 2 vols., illustrated.

Snake Bite Kits and Antivenin

MANY INQUIRIES about snakebite kits and where they may be obtained have been received at the museum; so it seemed advisable to list here the Canadian sources of supply known to me.

Antivenin for rattlesnake bite may be obtained from John Wyeth and Brother (Canada) Limited, Walkerville, Ontario, or from the Toronto office of this company at 533a Eglinton Avenue West, or through any druggist, as mentioned above. The antivenin kit includes a hypodermic syringe, which, after the injection, may be used as a suction device.

The Cutter Compak Suction Snake-bite Kit may be obtained from the Cutter Laboratories International, Calgary, Alberta (Ontario sales: Donald Wilson, 24 Courton Dr., Scarborough; Earl H. Maynard Co., 1619 Weston Rd., Weston; Standard Hospital Supply Ltd., 20 Belivia Rd., Etobicoke). It contains also an incising blade, a constricting cord and a vial of antiseptic. The entire kit measures only 3×1 inch and weighs 1¼ ounces; so it may be carried in the vest pocket or purse. This is an excellent small kit, and inexpensive. The Cutter Laboratories also stock antivenin.

A splendid kit, more elaborate than the preceding, is supplied by the Mine Safety Appliances Co. of Canada Ltd., 148 Norfinch Dr., Downsview, Ontario (branches at Calgary, Edmonton, Halifax, Montreal, Saskatoon, Sydney, Winnipeg, Vancouver). It is enclosed in a metal case, $4 \times 2 \times 1$ inches, and contains the following: standard tourniquet with spring buckle fastening, antiseptic brush, incising knife, self-suction pump with adapter for fingers and toes, ammonia inhalants and adhesive compresses.

Information about other pertinent equipment carried by these companies may be obtained from them upon request.

For the Attending Physician and Hospital Staff

ALTHOUGH as a herpetologist I have given some study to the subject of snakebite, its effects and treatment, I am not a physician and profess no claim to medical knowledge. However, I was singularly fortunate in obtaining the generous co-operation of Dr. Robert I. Harris of Toronto as medical consultant in the preparation of the following notes addressed to the medical profession.

In view of the fact that snakebite is rare in Canada, and, consequently, few doctors here have had experience with it, a few remarks on medical procedures used in the United States, and elsewhere, where this kind of accident is more common, would seem to be in order; also some notes on the nature and action of snake venoms, since they bear directly upon the treatment.

There is a considerable literature, but few in Canada have had occasion to consult it, and there is not time to do so in an emergency. What is epitomized below is brief and for emergency purposes only, but is from reliable sources.

NATURE OF VENOMS. It is generally known that there are two broad, natural groups of snake venoms, the neurotoxic and haemotoxic, and that viper venoms in general belong to the haemotoxic group. It is less generally known that the two groups of venoms overlap in their actions, each group containing factors which are characteristic of the other. Thus, viper venoms, although mainly haemotoxic, *do contain neurotoxins* in sufficient quantities to be extremely dangerous. This is very important, because these insidious elements may assert themselves after the more spectacular haemotoxic symptoms appear to be well under control.

The venom of every species of snake differs from that of every other, but in general, rattlesnake and other viper venoms are rich in haemorrhagins, cytolysins, haemolysins, digestive ferments, thrombase, antifibrins, anticoagulins and antibactericidins. The toxic factors themselves are high molecular weight proteins. In viper venoms, generally, these complex molecules are too large to dialyze through the walls of blood vessels and enter the blood stream directly; so they are carried at first in the lymph. This is not true of some neurotoxic venoms, and in the bites of certain (perhaps all) cobras, and probably coral snakes, the use of a tourniquet, or other means of retarding the lymph flow from the bitten limb, is of doubtful, if any, value.

EFFECTS OF RATTLESNAKE BITE. The bite of a rattlesnake is usually, but not invariably, characterized by severe local pain. One or two fang punctures will be evident. Extravasation of tissue fluid is rapid at the site, with local oedema, swelling and purplish discolouration following soon. This advances up the bitten limb, and if not arrested, may reach and even invade the trunk. Some neurotoxic symptoms may appear soon, as tingling of the limbs, numbness of the face and lips, partial paralysis of the tongue and throat, difficulty in swallowing, vomiting, difficulty in breathing and cardiac pain. As more venom seeps into the general circulation, respiratory and cardiac distress may become acute. There may be extensive destruction of red corpuscles, leukocytes, capillary walls and muscle tissue, with haemorrhage from various internal organs and bleeding from the mucous membranes of the mouth and nasal passages. Shock usually develops quickly and may be very severe. The neurotoxic factors exhibit a special affinity for the control centres and nerves regulating respiration and heart action.

ANTIVENIN. In the section on first aid procedure recommended for the layman it was advised that the administration of antivenin should be left for the doctor, if one can be reached within three or four hours after the bite. It should, nevertheless, be used as soon as possible, but has saved life even after coma has set in. In severely poisoned patients antivenin treatment may have to be continued for several days. When venom has been delivered directly into a vein, large intravenous doses of 70 cc. or more must be used immediately. Complete directions accompany each package of Wyeth Antivenin (Nearctic Crotalidae) Polyvalent.

The Ontario Department of Health has established antivenin depots at twenty-five places in Ontario. A list of them will be found on page 85 in the section on first aid.

CONTROLLING THE SPREAD OF THE VENOM. Field tourniquets should be inspected when the patient arrives at the hospital or doctor's office, and readjusted if too tight or too loose. The tourniquet should be left on until after antivenin has been administered. It may then be removed, but the bitten limb should be immobilized, which may be done with a splint or plaster (with due regard for the inevitable swelling). The value of this lies in the fact that there is no lymph flow from an immobilized limb, so that the spread of the haemotoxic factors, at least, may thus be retarded or stopped.

REMOVAL OF VENOM FROM REGION OF BITE. It has been estimated by some authorities that as much as half of the total dose of venom received may sometimes be removed by incision and suction within the first few hours. This is important when it is realized that one or two drops more or less might decide the fate in any borderline case. Incision should be done under a local anaesthetic. The injection of a few hundred cc. of 1 per cent salt solution into the tissues around the bite, followed by continued suction, greatly assists in the rapid removal of venom.

Other details regarding incision and drainage are given in the first aid section.

Experiments on dogs conducted by Pope and Peterson indicate that negative pressure alternating between 60 and 120 mm. Hg may be the most effective means of removing venom. In human beings this pressure could be accomplished by the use of a Pavex boot.

BLOOD TYPING AND PERIODICAL EXAMINATION. The patient's blood should be typed as early as possible, a transfusion may be required later. A periodic (daily or oftener) examination of the blood and its chemistry, including red cell counts, haemoglobin and urea estimates, and detection of free haemoglobin in the serum, should be continued until recovery is complete. Severe anaemia which sometimes follows rattlesnake bite should be watched for and treated at the discretion of the physician.

KIDNEY FUNCTION. The kidneys are highly susceptible to damage after rattlesnake bite. A careful watch should be kept on the quantitative output and qualitative nature of the urine. This should include daily (or more frequent) microscopic and chemical analysis for the detection of erythrocytes, casts, free haemoglobin, non-protein nitrogen, albumin or other abnormal contents that would indicate a failure of kidney function, or the progression of haemolysis or cytolysis elsewhere in the body. Intermediate nephron nephrosis (resulting from haemolysis and muscle autolysis) commonly occurs after two or three days in severe snakebite poisoning, and death may result from kidney failure.

RELIEF OF PAIN. In rattlesnake bite pain may be severe and prolonged, the use of such drugs as morphine or barbiturates is therefore standard procedure. In this connection, certain new drugs, ACTH and cortisone, appear to be very effective. Further reference will be made to them under the heading "New Drugs."

CONTROL OF INFECTION. After rattlesnake bite the natural resistance of the body to pathogenic organisms is reduced practically to zero. The breakdown of tissues caused by the destructive action of the venom, coupled with its antibactericidal effects, greatly predisposes to infection. The danger is further enhanced by the presence in the mouths of some rattlesnakes of the causative agents of tetanus and gas gangrene. Indeed, infection rather than the direct action of the venom is sometimes the cause of mutilation or death. It is therefore standard practice to administer specific injections against tetanus and gas gangrene, and also penicillin or other antibiotics at the discretion of the physician.

REPLACEMENT OF LOST FLUID. The loss of fluid from the blood through extravasation may be considerable. Give fluids freely by mouth, as much as the patient desires. In addition to this, intravenous injection of normal saline with glucose or dextrose, Ringer's solution, or blood plasma, is often necessary and always beneficial.

BLOOD TRANSFUSION. Where haemolysis has been severe—which is not uncommon—or where there has been much loss of blood through haemorrhage, whole blood transfusions may be necessary to save life, or at least to promote recovery. The need in each case will have to be assessed and decided by the attending physician. The importance of blood typing, as mentioned above, as early as possible in every case, before the enzymic action of the venom has had time to vitiate the picture, cannot be overstressed.

NEW DRUGS. Certain of the new drugs, ACTH, cortisone and antihistamines, are beginning to attract attention as helpful agents in the treatment of snakebite poisoning. There have been tests on experimental animals and trials on a few human patients; the reports on some of the animal tests, and on some human beings, were encouraging. At this point, I can do no better than to quote, with Dr. Harris's kind permission, from his article in the *Canadian Medical Association Journal* (1957, pp. 874–8): "Cortisone and ACTH may prove to be the most valuable of all agents in the management of rattlesnake bites. Experience as yet is too meagre to be certain of their action, but their well-established merit in blocking the haemolytic reaction of incompatible blood transfusion and in the relief of asthma makes it possible that they can block the enzymic reactions of snake venoms. There is some experimental evidence to support this. The first dose of cortisone or ACTH should be given intravenously to ensure the most rapid effect. After that they may be administered by mouth."

WARNING SIGNS. Haemoglobin or blood in the urine or stools, increased pulse rate, falling blood pressure, respiratory or cardiac distress, the appearance of ecchymoses remote from the region of the bite, or any tendency to bleed from the mucous membranes: the presence of one or more of these symptoms indicates that venom is still active in the body and suggests continued treatment with antivenin, along with whatever other measures the attending physician may consider advisable. As mentioned earlier, a constant vigil should be maintained for the delayed onset of neurotoxic symptoms, even after recovery appears to be assured.

Additional Note on the Use of Cortisone and Local Medical Facilities

By DR. R. I. HARRIS

People who summer in Georgian Bay and the Bruce Peninsula should know that several of the doctors who spend their holidays in Georgian Bay have equipped themselves to deal with rattlesnake bites. Summer residents in Georgian Bay should make themselves familiar with the facilities for treatment which may be available in their own community. When such local facilities exist in charge of a doctor, valuable time often can be saved in the

90

initiation of emergency treatment (the administration of antivenin and cortisone), after which the victim can be transported to the nearest hospital for continuation of treatment.

There is evidence which suggests that cortisone is valuable in the treatment of rattlesnake bites. Research to determine its merit is in progress. Until this research can be completed, it would be wise to administer a cortisone preparation in every case of rattlesnake bite. A preparation which can be administered intravenously, such as Solu-Cortef (Upjohn) is best and it should be given at the earliest possible moment. It can do no harm and it may do much good. It can be administered while the antivenin is being prepared and the tests for sensitivity carried out.

Medical and Technical References

AMORIM, MOACYR F., and MELLO, RAUL F. Intermediate Nephron Nephrosis from Snake Poisoning in Man. *Am. J. Path. 30* (3) (whole no. 178): 479–499 (1954).

BARNES, J. M., and TRUETA, J. Absorption of Bacteria, Toxins and Snake Venoms from the Tissues. *Lancet,* pp. 623–626 (snake venoms, pp. 623–624) (1941).

BUCKLEY, ELEANOR E. *Antivenin (Nearctic Crotalidae) Polyvalent.* John Wyeth & Brother Limited, 1947. Pp. 16.

CLUXTON, H. H., Jr. The Treatment of the Black Widow Spider Bite and Copperhead Snake Bite with ACTH. *Proc. 2nd Clin. ACTH Con. 2:* p. 445.

GOWDY, JOHN M. (Wm. W. Hastings Indian Hosp., Tahlequah, Okla.). Treatment of Snakebite with Cortisone: Report of Four Cases. *Am. Prac. & Dig. Treat. 5:* 569–572 (1954).

HARRIS, ROBERT I. Rattlesnake Bites. *Canad. M. A. J. 76:* 874–878 (1957).

HOBACK, WILLIAM W., and GREEN, THOMAS W. (Clinch Valley Clin. Hosp.). Treatment of Snake Venom Poisoning with Cortisone and Corticotropin. *J. A. M. A. 152:* 236–237 (1953).

JAFFE, FREDERICK A. A Fatal Case of Snake Bite. *Canad. M. A. J. 76:* 641–643 (1957).

KLAUBER, LAURENCE M. *Rattlesnakes: Their Habits, Life Histories, and Influence on Mankind,* 2 vols. Univer. Calif. Press, 1956. Pp. xxix, 1476. ("The Bite and Its Effects," pp. 797–859; "Treatment and Prevention of the Bite," pp. 860–972).

KNAPP, WILLIAM A., and FLOWERS, HERSCHEL H. (Sch. Vet. Med., Univer.

Georgia). Treatment of Poisonous Snakebite in the Dog with Cortisone Acetate. *Vet. Med. 51*: 475–478 (1956).

POPE, CLIFFORD H., and PETERSON, L. W. Treatment of Poisoning with Rattlesnake Venom: Experiments with Negative Pressure, Tourniquet and Bulb Suction. *Arch. Surg. 53*: 564–569 (1946).

STICKEL, WILLIAM H. *Venomous Snakes of the United States and Treatment of Their Bites*. U.S. Dept. Interior, Fish and Wildlife Service, Leaflet 339. 1952. Pp. 1–29.

YOUNG, NETTIE. Snakebite: Treatment and Nursing Care. *Am. J. Nursing 40* (6): 657–660 (1940).

INDEX

CANADIAN UNIVERSITY PAPERBOOKS

Other titles in the series

Milton Keynes UK
Ingram Content Group UK Ltd.
UKHW011831100924
448154UK00010B/156